Graphics, Touch, Sound and USB

User Interface Design for Embedded Applications

Lucio Di Jasio

All rights reserved
No part of this book may be reproduced, scanned, or distributed
in any printed or electronic form.

Copyright© 2014 by Lucio Di Jasio

Visit us at: http://www.flyingpic24.com

E-mail: pilot@flyingpic24.com

An acknowledgment and thanks to Lulu Enterprises, Inc. for making the
publishing of this book possible.
http://www.lulu.com

ISBN: 978-1-304—60654-9

Printed in the United States of America

Second Edition: 2014

This book is dedicated –

To Sara and Luca

Table of Contents

Preface 1

Introduction 3
    Software Choices 3
    Hardware Choices 4
    What This Book is Not 5
    Online Support 5
    Second Edition Notes - October 2014 5

Chapter 1 7
Hello World 7
    About MPLAB X 7
    Installing MPLAB X 10
    MPLAB XC16 11
    Installing MPLAB XC16 12
    MPLAB XC16 Under the Hood 13
    The PIC24 Mikromedia Board 16
    Choosing Your Programming Tool 17
        Mikromedia Bootloader 17
        PICKit™3 – In Circuit Programmer and Debugger 18
        mikroProg – In Circuit Programmer and Debugger 18
        ICD3 and REAL ICE – In Circuit Programmers and Debuggers 18
    Let's Get Coding 19
        The Working Directory 19
        New Project Checklist 19
        MPLAB X Projects Under the Hood 20
        Logical Folders 21
        New File Checklist 23
    Hello Embedded World 24
        Mikromedia Bootloader Notes 27
        Microchip HID Bootloader Notes 27
        Finding the Executable File 28
        Mikromedia Bootloader Programming Steps 28
        PICKit3, ICD3 or REAL ICE Programming Steps 28
        mikroProg Programming Steps 28
    Testing Hello World 29
        Mystery Solved 30
        Configuring the PIC24 30
    Summary 31
    Tips & Tricks 32
    Suggested Reading 32
    Exercises 33
        Solution 34

Chapter 2 35
Hello MLA 35

Preparation..........35
Getting the "right" MLA..........35
    Current MLA vs. Legacy MLA..........36
    Downloading the MLA..........37
    Installing the MLA..........37
    Managing MLA revisions..........38
    MLA Under the Hood..........39
    Inside the Microchip Folder..........40
    A Bit of Structure in the Working Directory..........41
The Graphics Library..........43
    Configuring the MLA..........44
    Creating a New Project..........44
    Creating the Hardware Profile..........45
    Creating PICconfig.h..........46
    Graphics Library Driver and Primitives Checklist..........47
    Adding the Display section of the Hardware Profile..........49
    Customizing the Graphics Configuration file..........51
    GREEN Light..........53
    Final Checks..........54
    Build Failure..........55
    Setting the Include Path..........57
    Second Attempt..........58
Playing with the Primitives..........59
Finally, Hello MLA!..........60
Tips & Tricks..........61
    Code Browsing..........61
    Code Completion..........62
Summary..........62
Suggested Reading..........62
Exercises..........63
    1- Centering text on the display..........63
    2- Fading In..........63
    Solution 1..........64
    Solution 2..........65

Chapter 3..........**67**

Graphic Resources..........**67**
    Graphic Resources and Performance..........67
    Alternative Microcontroller Choices..........68
    Preparation..........69
Bitmaps..........69
    Primitive Bitmap Support..........70
    The Graphic Resource Converter..........71
    Bitmap Files Under the Hood..........75
Fonts..........77
First Demo Project: Splash Screen..........80
    Going Large!..........81
    A Colorful Splash Screen..........82
Terminal Emulation..........83

About Scrolling Text.. 83
Sharing the Terminal Module.. 92
Summary.. *92*
Tips & Tricks.. *93*
Online Resources *94*
Suggested Reading... *94*
Exercises.. *95*
Solution to Exercise 2... 96

Chapter 4..97

Touch Input...97
Touch Technologies.. 97
Elements of a Resistive Touch Screen.. 98
Board Support Package – TouchScreen.c 100
Configuring the Touchscreen.. *100*
First Touch... *103*
Preparation... 103
Touch Demo... 104
Calibration.. 106
Touching What... 107
On a Grid... 109
Creating the TouchGrid module.. 111
A Grid Demo.. 112
Summary... *115*
Tips & Tricks... *116*
Skipping Calibration... 116
Suggested Reading... *116*
Online Resources *117*
Exercises.. *117*
Solutions... 118

Chapter 5..121

Storage...121
Accessing a File System.. *121*
Introducing FSconfig.h.. 123
Introducing uMedia.c... 125
Reading Text Files.. *127*
Preparation... 127
Repetita Juvant or does it?.. 128
Back on track... 129
Selecting a File from a List... *131*
Menus.. 132
Reading Image Files.. *137*
Image Decoders.. 137
Creating LCDmenu.. 138
Configuring the Image Decoder.. 141
Displaying Slides... 142
Choosing the Right File Format for your Images.. 143
Serial Flash.. *145*

A Serial Flash Demo Project *147*
    Preparation 147
    HardwareProfile.h Serial Flash section 147
Tips & Tricks *151*
    Using Serial Flash to store TouchScreen Calibration Data 151
Summary *152*
Suggested Reading *152*
Exercises *152*
Solution to Exercise 3 .... *153*
Online Resources *154*

Chapter 6 155

Sound 155

    The VS1053 codec 155
    Serial Command Interface 157
    VS1053.h 159
    Hardware Profile VS1053 Decoder Section 160
    Serial Data Interface and Protocol 160
    Preparation 161
    Initialization and Serial Communication 162
    Sine Test 164
Hello Again! *167*
    Creating VS1053.c 168
Playing Audio Files *173*
Tips and Tricks *176*
Summary *177*
Suggested Reading *177*
Online Resources *177*
Exercises *178*

Chapter 7 179

Graphics Object Layer 179

    An Overview of GOL 179
    Widgets and Objects 180
    Blocking vs. Non-Blocking 181
    Object States 182
    About Style 182
    Creating a Widget 184
    Messaging 186
    The Active Object List 189
    Drawing 190
A First GOL (Application) *191*
    Preparation 191
    GOLSimple Testing 194
A Slider Example *194*
    Creating a Window 196
    Defining A Color Scheme 197
    GOLMsgCallback, Light, Action! 198
Graphics Display Designer *200*

GDD X Project Preparation... 200
Understanding GDD X Code... 203
Adding User Code to a GDD X Generated Project... 206
GDD X Checklist.. 208
Summary... *209*
Tips & Tricks... *210*
Editing the GDD templates.. 210
Adding Support for a New Board... 210
Online Resources... *210*
Suggested Reading... *210*
Exercises... *211*
Solutions.. 212

Chapter 8 215

USB..215

A Very Brief Introduction to USB... *215*
Coding with Class 216
The Drivers Problem.. 217
The Physical Layer... *218*
Power.. 219
Beginning from the End(point)... *219*
Types of Transfers.. 220
Enumeration.. 221
MLA USB support modules... 222
Communication Device Class.. *223*
CDC Descriptors... 223
Using the CDC class 228
Inspecting usb_config.h.. 229
Getting Attached... 231
Virtual Serial Ports.. 232
Puts and Gets... 234
The Callback Handler... 234
Using the Start of Frame Handler 236
Blocking I/O with the CDC Class 238
A CDC Touch Screen mini Terminal.. 239
CDC Summary... 242
USB-CDC Applications Development Checklist.. 243
Human Interface Device Class... *244*
Using the HID Class.. 247
Establishing the Connection... 249
HIDTxPacket() and HIDRxPacket()... 250
HID Simple Demo... 252
HID Summary.. 258
HIDAPI 258
Rapid Development with Python and HIDAPI 259
HID Applications Development Checklist.. 261
Summary... *262*
Tips & Tricks... *262*
Getting a VID/PID pair... 262

 Using multiple classes with Composite Devices 262
Suggested Reading *263*
Online Resources *263*
Exercises *263*

Preface

Against my best efforts even this book turned out to take way more pages and time than initially planned. It would have never been possible for me to complete it if I did not have 110% support and understanding from my wife. Special thanks also go to Steve Bowling, a friend, a pilot and an expert of all Microchip things, for reviewing the technical content of this book and providing many helpful suggestions for the demonstration projects and hardware experiments. As usual I owe big thanks to Vince Sheard for patiently listening to my frequent laments, always working hard on making MPLAB® a better tool and Rawin Rojvanit and his entire team for addressing quickly all my questions and offering so much help and insight into the inner workings of the Microchip Libraries for Applications.

Special thanks go to Nebojsa Matic and his wonderful team at MikroElektronika as they keep amazing me with innovative ideas and quality products.

Once more I would like to extend my gratitude to all my friends, colleagues at Microchip Technology and the many embedded control engineers I have been honored to work with over the years. You have so profoundly influenced my work and shaped my experience in the fantastic world of embedded control.

Finally thanks to all my readers, especially those who wrote to report ideas, typos, bugs, or simply asked for a suggestion. It is partially your "fault" if I embarked in this new project, please keep your emails coming!

Introduction

"The story goes that I wanted to write a book about flying..."
Ok, that was the beginning of my first book, "Programming 16-bit Microcontrollers in C", aka the "Flying PIC24" book as I like to call it because of the many references I made to the world of aviation. Some readers loved them, some could not appreciate the *detours*, so by the second book, "Programming 32-bit Microcontrollers in C", I dropped the *aviation* theme in favor of a simpler *exploration* of 32-bit microcontrollers programming. Regardless, the objective of both books was to introduce C Programming, as it applies to Embedded Applications, and I assumed that the reader had either no previous experience programming an embedded microcontroller or was an experienced *assembly* programmer just recently confronted with the challenge of using a higher level programming language.

In this book, I am going to take for granted your C programming skills, basic knowledge of use of 16 and 32-bit microcontrollers and their peripherals (I/Os, TIMERS, PWM, SPI and I2C) and I am going to progress rapidly on to more advanced subjects that are central to the design of User Interfaces for Embedded Applications.
You will learn how to interface to color *graphics* displays (TFT) with *touch* screen inputs to design compelling graphical user interfaces. You will use *sound* to get the attention of the user and/or provide quality audible feedback. You will store and retrieve data (fonts, audio, images...) from S*erial Flash* devices and *microSD* cards and we will expand our connectivity options to include *Full Speed USB* to communicate with personal computers and other devices.

Software Choices

Obviously we won't be able to develop all the code required from scratch. We will focus instead on getting the most out of the **Microchip Libraries for Applications or MLA**. This is a professionally written and maintained set of libraries for graphics and connectivity solutions that Microchip Technology, Inc. makes available (royalty free[1]) to all PIC® microcontrollers users and constitutes an excellent starting point for every embedded control designer.
Since the MLA is supplied entirely as C source code, although we won't be able to claim of having *written* every single line of code, at least we will be able to say that we *read* it all. Even more interestingly, the MLA is architecture agnostic, as most/all of the included libraries can be compiled for any of the 8-bit, 16-bit and 32-bit PIC architectures!

---
[1] Microchip MLA license imposes a single limitation in the use of the library source code: you must use it on PIC microcontroller!

Since we will use **MPLAB X**, a cross platform (Windows, OS X and Linux) integrated development environment (IDE), and the unified **MPLAB XC compiler suite**, all the lessons learned throughout the book will be immediately applicable to a large selection of the 1000+ microcontrollers models supported as of this writing.

Please note that **the MLA is a living project** and as such it could and will eventually diverge from the version I used while writing this book (June 2013). Please **make sure to download the correct version of the library** when proceeding through the initial set up as detailed in Chapter 1.

Hardware Choices

Since this is meant to be a hands on learning experience, I chose to base all the demo projects in this book on a specific hardware platform, the ***PIC24 Mikromedia*** board, a product of MikroElektronika (http://www.mikroe.com).

This is a very well engineered little board that is barely covering the back of a 3.2" TFT display, packed with all the features needed to build a *quasi-multimedia* user interface, but inexpensive enough to fit the tightest budgets.

In fact Mikromedia boards are often purchased in volume and mounted directly in *industrial* application as an advanced display module or as the main control board of small embedded systems. By visiting the MikroElektronika web site you will discover that there is a whole family of boards featuring a large number of 8, 16 and 32-bit microcontrollers.

My decision to go with the PIC24 Mikromedia model is based on convenience and simplicity. The device performance is more than sufficient for even the most advanced examples featured in this book while the architecture is simple and does not get in the way as we focus on the details of the user interface design. Porting the code and examples to smaller (PIC18 Mikromedia) or larger and faster (PIC24E, dsPIC and PIC32 Mikromedia) models will require only minor changes, mostly relegated to the *Hardware Profile* and the *Board Support Package* folder.

While Mikromedia boards are well documented and are supported by the MikroElektronika own toolchain, including In Circuit Programmers, Debuggers, and C, Basic and Pascal compilers, in this book I chose to stick to the **standard Microchip toolchain** for maximum portability.

My debugger/programmer of choice has been the **PICKit™ 3**, which I highly recommend for its versatility and low cost. If you can splurge (doubling the board budget) go for the ICD3 instead. Both will give you pretty much the same functionality and access to any PIC microcontroller model, but the ICD3 will be considerably faster. This is not going to be an issue with most of the small examples in the book, but the difference might start to show when we will be flexing the MLA muscle filling a large portion of the microcontroller flash memory with code and data.

What This Book is Not

This book is not an introduction to Embedded Programming, or a primer in C programming. As I mentioned before, this book assumes already a *medium* level of expertise in the use of microcontrollers, preferably but not necessarily PIC microcontrollers.

The complexity of the projects developed increases steadily throughout the book, and in all cases require a certain agility in handling multiple source files and relatively deep nested include files. This book does not replace the microcontroller datasheet or the board user manual and documentation, in fact I will often refer you to such material for further study. Similarly this book cannot represent a comprehensive summary of all the features offered by the Microchip Libraries for Applications and it does not replace the extensive documentation available for each of its modules.

Should you notice a conflict between my narration and the official documentation, ALWAYS refer to the latter. However, when you do so, please remember to send me an email, I will publish and share any correction and/or useful hint on the blog and book web site.

Online Support

All the source code developed in this book is made available to all readers on the book web site at: http://www.flyingpic24.com This includes additional fonts, images and audio resources together with a complete set of links to online code repositories and third party tools as required and/or recommended in the book.

Over the last few years I have been contributing to a blog, "The pilot logbook", at http://www.flyingpic24.com/blog and I will continue to do so time permitting.

Second Edition Notes - October 2014

This second edition follows the first by mere six months. This would be an unusually short period of time for a traditional book but, given the extreme flexibility of self publishing, this is a great opportunity available to me (the author) to ensure that you get the most value for your money.

In this second edition I have already taken care of all Errata known to date and I took the opportunity to clean up and streamline the presentation in two important chapters related to Touch and Sound. Minor aesthetic improvements have also been added to a few of the demo projects to enhance their visual impact, hopefully without compromising the clarity of the presentation.

Chapter 1

Hello World

In this chapter we will prepare our initial software and hardware setup. This will include, downloading and installing:

- *MPLAB® X* (v1.90 or later), this is the integrated development environment that will allow us to organize projects, edit, program and debug. MPLAB X is common to all PIC® architectures
- *MPLAB XC16* (v1.21 or later), this C compiler is common to all PIC24 and all dsPIC® models

We will gather some critical pieces of documentation and begin inspecting the structure of an MPLAB X project so to be able to understand how to best maintain the code over the next few chapters.

We will also start reviewing some basic hardware elements of the PIC24 Mikromedia board and, in the process, we will put it to test to check if our toolchain is ready and the board is alive!

NOTE

> In this first chapter we will add a few extra initial steps, that won't be repeated necessarily in the following, to ensure the correct installation of all our tools before we start creating the first project. I suggest that you follow these steps with extra patience and accuracy, as failing to do so might result in significant waste of time and frustration later.

About MPLAB X

There is no doubt that every single exercise in this book can be launched and debugged with even the oldest of MPLAB versions (as long as you can find the PIC24FJ256GB110 in the supported devices list) but you are doing yourself a great disservice if you don't take advantage of the new MPLAB X IDE for a long list of reasons, among which, I would include the following few vital ones:

- MPLAB X runs on Windows, OS X and Linux smoothly and reliably. I have been using the OS X version since the very first betas and I have developed and tested each and every project of this book on Windows and Mac. The OS X and Linux versions are not just *ports* of the IDE, they work just as well as the Windows version and often better/faster.

- MPLAB X has far superior code browsing capabilities. Some of these capabilities (see the *Navigate to* option in the editor context menu for an example) can save you long hours of tedious research when trying to understand the inner workings of a complex library (and the Microchip Library of Application can seem quite an intricate labyrinth at times).

- MPLAB X includes support for *versioning* (that is the ability to keep track of changes in your code and preserve a complete history and back up of all your work) interfacing with Subversion and Mercurial. This is essential stuff if you work in a team, but even if you don't, you will appreciate how MPLAB X includes its own basic versioning system of sorts in the form of the *Local History* tool. Once more, this can save you a lot of time and frustration in the future as you will see how using libraries, with the growing number of files in our projects, will make it harder for us to remember what we changed, where and why.

- MPLAB X projects can be made truly *relocatable*. This means that if you take some basic precautions you will be able to move the entire project directory on your hard drive without destroying or damaging the project functionality. This is very important in larger applications where you might be restructuring your code, versioning it, and in general migrating it from place to place or to different machines over time.

- And finally, eventually Microchip is going to phase out MPLAB 8. At the time of this writing (August 2014) new releases of the Microchip Libraries for Applications do not offer anymore MPLAB 8 project examples!

Title	Date Published	Size	D/L
Windows (x86/x64)			
MPLAB® X IDE v2.20	9/3/2014	377.6 Mb	
MPLAB® X IDE Release Notes / User' Guide v2.20 (supersedes info in installer)	9/3/2014	4.0Kb	
MPLAB® X IDE Chinese Translation Files v.1.80	08/08/2013	22Mb	
Linux 32-Bit and Linux 64-Bit (Requires 32-Bit Compatibility Libraries)			
MPLAB® X IDE v2.20	9/3/2014	345.3Mb	
MPLAB® X IDE Release Notes / User' Guide v2.20 (supersedes info in installer)	9/3/2014	4.0Kb	
MPLAB® X IDE Chinese Translation Files v.1.80	08/08/2013	22Mb	
Mac (10.X)			
MPLAB® X IDE v2.20	9/3/2014	254.9Mb	
MPLAB® X IDE Release Notes / User' Guide v2.20 (supersedes info in installer)	9/3/2014	4.0Kb	
MPLAB® X IDE Chinese Translation Files v.1.80	08/08/2013	22Mb	

igure 1.1 - MPLAB X Download page

Don't procrastinate any longer, get MPLAB X (latest release you can download) and start up the learning curve. I will show you some of the tricks along the way, but you will have to put some time into it no matter what.

ONLINE RESOURCE

Use this shortcut to go straight to the MPLAB X support page:
http://www.microchip.com/mplabx

Installing MPLAB X

Here is what's important to remember when you install MPLAB X:

1. Forgive the obviousness, but make sure to install the right version for your OS. Some of the links on Microchip web site are *smart,* others require you to choose depending on the page you landed on. Check your browser download window and verify that you are getting an *.exe* file if you are aiming at Windows, a *.dmg* file for OS X and a *tar.gz* file for Linux. There is nothing more frustrating than waiting for a 200-300 MByte download taking several minutes (or hours depending on your internet connection) only to realize you downloaded the wrong thing.

2. If you had a previous version of MPLAB X installed, remember that you have to uninstall it first! In case you were wondering, yes, you can trust MPLAB X un-installer. Your projects, tools and configurations will be remembered and will be restored safely as soon as the new version is up and running. On the contrary, there is no need to uninstall previous versions of MPLAB 8. MPLAB X will NOT overwrite or corrupt previous MPLAB 8 installations. The two tools can co-exist peacefully on the same machine. In fact you might want to keep both, side by side, to handle old and new projects.

3. Now it's time to launch the installer and, after the usual click-through license steps, you will be asked to choose (or confirm a proposed) installation directory. Depending on the operating system there are some obvious choices but you always have a degree of freedom. For example:

   - In Windows you will be offered to install MPLAB inside the usual path: "**C:\Program Files**" so that it will be easy to reach via the "**Start**" menu

   - On a Mac, it will be a folder on the path: "**/Macintosh HD/Applications**" so that it will be accessible from the dock "**Apps**" folder.

   In all cases a folder called **Microchip** will be created and inside it another folder called **MPLAB X** will be nested. My recommendation is that you trust the tool. By choosing a different installation path, you might run the risk of confusing the installers that will follow as they might have some preconceived notion of where *they* should find MPLAB X and vice versa.

ONLINE RESOURCE

> Microchip Technical Training Team has developed a relatively large *wiki* (http://microchip.wikidot.com) that includes many short step-by-step tutorials and a good number of videos too. It's well worth browsing through it to check out the FAQs and some nifty tips and tricks.

MPLAB XC16

Contrary to MPLAB X IDE, a tool that covers all (one thousand plus) PIC® microcontrollers, MPLAB XC16 is specific to 16-bit PIC microcontrollers. This means that this compiler supports *only* PIC24, dsPIC30 and dsPIC33 devices, which still makes up for a long list of part numbers, running up to several hundreds. Most importantly not all MPLAB XC16 versions are 100% free. In fact much like its predecessor MPLAB C30, MPLAB XC16 is offered in various versions that offer more or less optimization options. They are:

- *Free version*, which offers only a basic set of optimizations choices: **-O0** and **-O1**
- *Standard version*, which offers a more complete set of optimization options, including space **-Os** and procedural abstraction **-O2**
- *PRO version*, which includes all of the above and the non-plus ultra mode **-O3**.

Both Standard and Pro can be had with a sort of floating license option (installing a *Network License Server*). This is an interesting proposition for large organizations where a large number of developers can share a conveniently small number of compiler licenses optimizing cost in addition to space and performance.

Note that the Free version includes also a (60 days) time limited PRO license. So you can have a taste of what those additional optimization modes can buy you and decide if the saving is worth the price based on the benefit to your specific application and coding style.

For the purpose of this book, as in my previous, we will NOT need any of the advanced features of the optimizer and the Free version will suffice to give us more than adequate performance. In fact, most applications will run perfectly satisfactorily without using even the most basic of the optimization levels (-O1)! Further, use of advanced optimization levels can make debugging code really hard and is generally discouraged during development.

NOTE FOR THE **MPLAB C30** EXPERTS

> In case you were wondering, the MPLAB XC16 compiler is pretty much based on the same engine of the previous MPLAB C30 compiler. There are minor differences though in the syntax of some of the *implementation-specific* extensions, such as pragmas, built-ins and configuration bit macros. These changes were introduced in order to standardize the compilers syntax among all PIC architectures and to smooth further the *migration* path.

Installing MPLAB XC16

From the same MPLAB X download web page (see Figure 1.1) you will find links to download the correct version of the compiler for your operating system.

After downloading a file of approximately 100Mbytes, launch the installer and you will be presented with a short sequence of dialog boxes:

1. A classic click-through license agreement. You **must accept** the legal terms and conditions if you want to continue.

2. You will be offered a choice between installing the XC16 compiler, a Network License server or updating your license settings. Choose to **install the XC16 compiler** at this time.

3. Next, you will be asked what kind of XC16 compiler license you plan on using. Unless you are part of a large organization where a Network Licensing scheme is in use, choose to **install the XC16 compiler on your computer** (local access key)

4. At this point you will be confronted for the first time with the *License Activation Manager,* which will prompt you to enter your license activation key. Unless you happen to have one such key, simply **leave the field empty** and **click on Next**.

5. Eventually **click on Yes** in the final confirmation dialog box (see fig. 1.2)

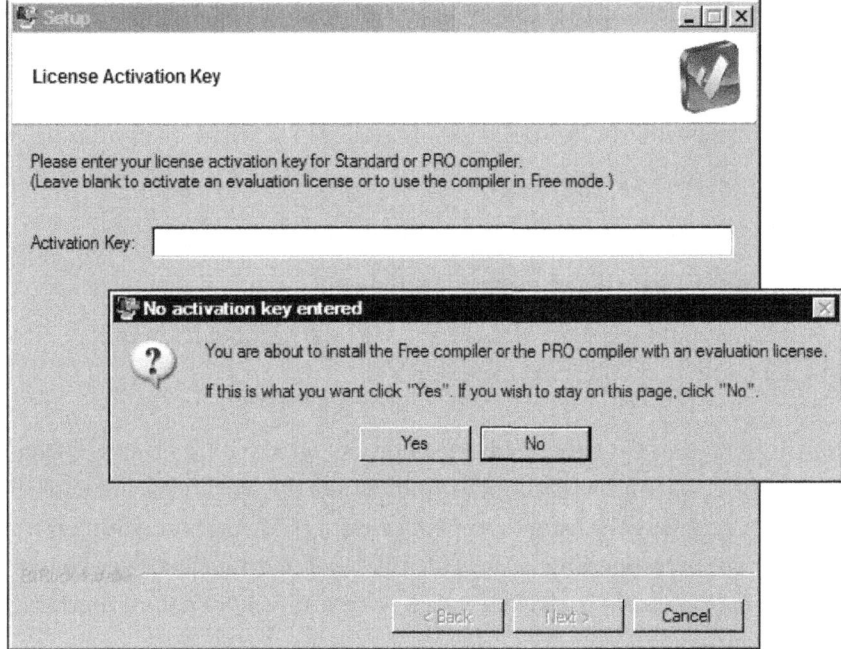

Figure 1.2 – License Activation Manager dialog

NOTE FOR THE **MPLAB C** LICENSE OWNERS

If you happen to own one or more MPLAB C30 compiler licenses, you can request and obtain from Microchip activation keys for equivalent XC16 licenses at no cost.

6. Lastly, you will be asked if you want to activate the time limited PRO evaluation period now. Note that this feature can be turned on at a later time, when most convenient to you. So don't worry about it for now, **run the compiler in Free mode**. Learn to use the tool first, develop your code. When ready, you will be able to launch the License Manager (XCLM tool) and start your 60 days evaluation period.

MPLAB XC16 Under the Hood

As soon as the XC16 installer terminates, take the time to explore the results of our installation thus far. Use your File Manager (aka Finder or feel free to type away at your Linux console) to navigate to the MPLAB X installation directory.

Start from the *Microchip* folder. In it you will find a number of sub-folders corresponding to each of the tools eventually installed. As a minimum, at this point you should find the following folders:

- *mplabx* – Here is where the IDE components are installed, including the main executable and all its plug-ins.
- *xc16* – This is were the compiler components have been placed
- *xclm* – The above mentioned License Manager

Additional sub-folders could include: *Mplab8.xx*, *MPLAB Cxx* older compilers, xc8 and/ or xc32 compilers depending on your past experiences.

We are mostly interested in the **xc16** folder for now, so let's drill in. You will find that within it there is (at least) another sub-folder indicating the specific revision of the compiler installed (*v1.21* was the latest as of this writing). If you had installed previous versions, there will be corresponding folders at this level. Notice that this scheme allows you to maintain numerous tools in parallel, each with multiple revisions, simultaneously available without conflict.

Keep drilling to reach another level of detail. Here you will find more sub-folders of interest to us, among which:

- *bin* – containing the actual compiler and linker executable
- *include* – containing the classic ANSI C header files (stdio.h....)
- *lib* – containing pre-compiled library modules required by the compiler run time (crt0...)
- *support* – containing the peripheral libraries and device specific header files, one for each 16-bit PIC and dsPIC model and one for each peripheral module
- *docs* – containing a most precious set of documents (see Figure 1.3)

Hello World - 15

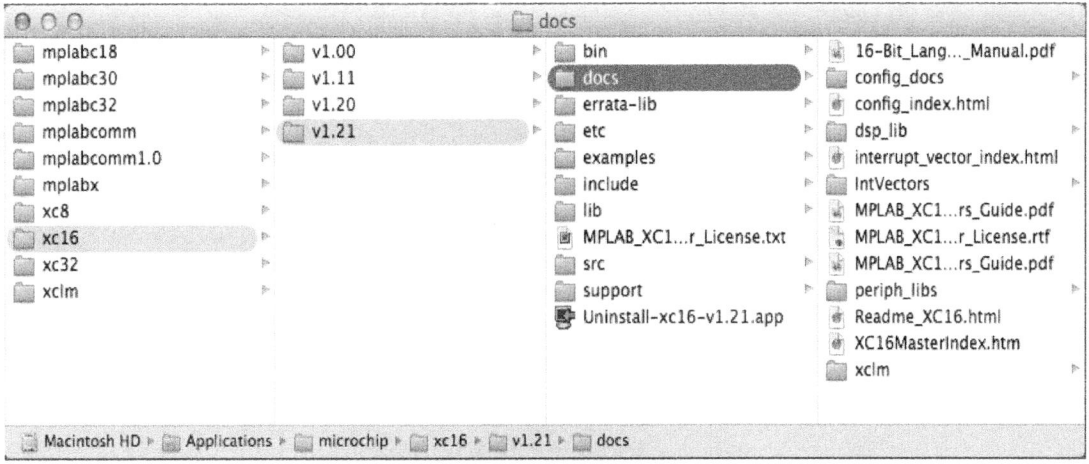

Figure 1.3 - xc16 folder contents

It is the *docs* folder that you will want to bookmark for later use. There are three key documents in here that you might want to access quickly and frequently:

1. *MPLAB XC16 Compiler User Guide*
2. *16-bit Language Tools and Libraries Guide*
3. *PIC24 Peripheral Libraries Guide* (inside the *periph_lib* subfolder)

These documents are all accessible in the portable acrobat (.pdf) format except for the very last one. If you are a Mac or Linux user though this isn't a big loss as you will find that browsing directly and automatically through the peripheral library header files (found in the *support/h* subfolder) is really easy and convenient thanks to the advanced source browsing features of MPLAB X.

Finally there is the X*C16MasterIndex*, an HTML file that will be quickly opened by your browser of choice. It provides a quick way to reach and navigate throughout all the documentation available.

Figure 1.4 - PIC24 Mikromedia Board

The PIC24 Mikromedia Board

The PIC24 Mikromedia board is part of a series of innovative *multimedia* development boards offered by MikroElektronika (notice the three *k*s), a long time third-party supplier of hardware and software tools for Microchip microcontrollers. Thanks to the deserved reputation for affordable, yet quality products, MikroElektronika has gathered a strong following in the educational and hobbyist market, with universities and technical schools from all over the world representing a strong and loyal customer base.

I chose to base the examples in this book on the PIC24 Mikromedia model in particular because:

- It features a PIC24 GB1 model, which with 256 Kbytes of flash and 16 Kbytes of RAM, represents an ideal extension of the PIC24 GA0 series featured in my previous introductory book. The GB1 series adds a few notable new peripherals, among which a USB serial interface and the flexibility of the Peripheral Pin Select (PPS)

- It ships with a convenient bootloader, which can reduce the total hardware investment to the sole board

- A 3.2" TFT display with QVGA resolution and 16-bit color with great contrast, LED backlighting and a resistive touch screen

- Two clock sources: an 8MHz system clock and a secondary 32kHz real time clock

- An audio codec capable of WAV, WMA, MP3, and Ogg Vorbis decoding with earphones amplified output
- A micro SD Card interface, for true mass storage (as in Giga Bytes)
- USB interface (mini device connector)
- A 3-axis, solid state accelerometer
- A serial flash memory for additional 8Mbit of storage
- Two rows of additional expansion connectors that can be used to mount various "shields" and interface to the real world

The documentation provided with the Mikromedia boards is quite well organized and complete, so I will not spend more time to tell you what's in there, but rather move on quickly to show you how.

Choosing Your Programming Tool

I will leave to you, the reader, the choice of the programming tool to use with the Mikromedia board. The vast majority of the material in this book will be, from now on, completely independent from it.
If you are looking for some guidelines, you will find the next section of interest.
Let's start with the lowest cost:

Mikromedia Bootloader

Pros: This is by far the cheapest option. No additional hardware is required, the USB cable comes in the box, each Mikromedia board ships with the bootloader preprogrammed and ready to go! A small graphical user interface is provided to allow re-programming of the board with a single mouse click.

Cons: Burn and Learn cycle. No true debugging capabilities, no breakpoints, no single stepping through the code. Also, as of this writing, only a Windows version of the GUI is available. (Check on book web site for a Python based portable version of the GUI for Mac and Linux)

Note that the default Mikromedia bootloader is **not** compatible with the Microchip HID bootloader (featured in Microchip literature).

Figure 1.5 - PICkit 3 In Circuit Debugger

PICKit™3 – In Circuit Programmer and Debugger
This is my recommended choice!
Pros: For a small additional amount of money, you can buy yourself a true universal programmer and debugger that you will be able to reuse with 1000+ additional PIC microcontrollers. It is integrated natively in MPLAB X and Microchip provides constant firmware updates and enhancements.
Cons: It is relatively slow, although you will notice this only on the largest of the project/examples when filling up most/all of the flash memory of a PIC.

mikroProg – In Circuit Programmer and Debugger
Pros: This is the obvious choice if you have already invested in other boards from MikroElektronika and possibly their compilers (BASIC, C and PASCAL) and their IDE.
Cons: By contrast, it is not directly supported by MPLAB X, although it is not difficult to run the two applications side by side (MPLAB X to build, MikroProg GUI to program)

ICD3 and REAL ICE – In Circuit Programmers and Debuggers
Pros: These two tools will offer the highest possible programming and debug speed. Most complete set of advanced debugging options. Directly supported by MCHP, natively integrated in MPLAB X.
Cons: They are relatively expensive by modern standards (USD 250-400). Will require you to double (or triple) the budget. Will also require an adapter (or ad hoc cable) to fit the 6-pin inline ICSP connector (the adapter is available via MikroElektronika online shop).

Hello World - 19

Let's Get Coding

This was quite a long intro for a typical chapter. If you have already used MPLAB X and the XC16 compiler, perhaps following my previous book "Programming 16-bit microcontrollers in C", you have probably browsed quickly through this part and are now itching to get your fingers on the keyboard.

The Working Directory

Allow me only one last preparatory step: let's create the *working directory* together!
I don't have any particular preference on *where* you should create it or what *name* you should use for it, but for the sake of clarity, from this moment on I will refer to it as the ***Mikromedia*** folder. I have created it as a subdirectory inside my main ***MPLAB X Projects*** directory, so that I can keep all the files that we will be using in this book neatly organized inside it. In its turn the *MPLAB X Projects* directory sits directly inside my user account (home) directory therefore it is part of a relatively short path:

~/*MPLABXProjects/Mikromedia*

(Windows users will replace the '~' symbol with 'C:' and use backslashes)

New Project Checklist

This simple step-by-step procedure is entirely driven by the MPLAB X *New Project* wizard.

From the **Start Page** of MPLAB X, select **Create New Project**, or simply select **File>New Project...** from the main menu it will guide us automatically through the following seven steps:

1. *Project type selection*: in the *Categories* panel, select the **Microchip Embedded** option. In the *Projects* panel, select **C/ASM Stand alone Project** and click **Next.**

2. *Device selection*: in the *Family* drop box, select **PIC24**. In the *Device* drop box, select **PIC24FJ256GB110,** and click **Next.**

3. *Header selection*: this step is skipped automatically, you don't need one with this PIC family

4. *Tool selection*: select **PICKit3,** or other supported programmer/debugger of your choice, and click **Next.**
   If you plan on using a bootloader or mikroProg programmer, don't worry about this selection, you might as well check the PICKit3 or select the simulator

5. *Compiler selection*: select **XC16**, and click **Next**.

6. *Plugin board selection*: this step is skipped automatically

7. *Project Name and Folder selection*: type "**1-HelloWorld**" (no spaces) as the project name, browse to your working directory and click **Finish** to complete the wizard setup.

After a brief moment, you will be presented with a new *Projects* window (see Figure 1.6). This will be empty except for a small number of *logical folders*.

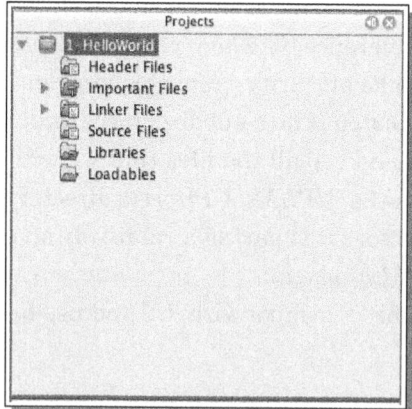

Figure 1.6 – MPLAB X Project Logical Folders

MPLAB X Projects Under the Hood

So what is a "project"? Let's take a look under the hood, as we explore, we are going to learn a few very useful things. Let's fire up the File Explorer (Finder for MAC users) and navigate to the working directory.

First of all you will find that MPLAB X has created a folder with the name you assigned to the project and appended the extension *.X* to it.

Think of this subdirectory as THE project!

NOTE FOR MPLAB 8 EXPERTS

> Since an MPLAB X project is effectively a folder, double clicking on it won't automatically launch MPLAB X, but simply will tell your file manager to inspect the directory contents. This is a sort of disappointment for old time MPLAB 8 users. If you are aching for some drag and drop action though, you can drag a project directory from your file manager into an MPLAB X (open) window.

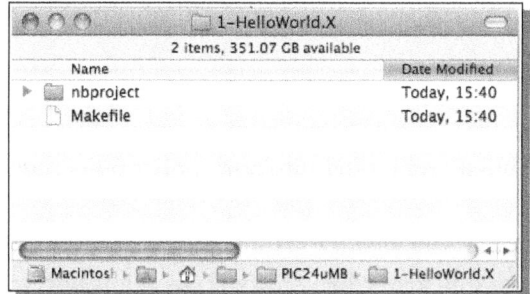

Figure 1.7 – MPLAB X Project Folders

You will notice how the project directory (see Figure 1.7) is not completely empty; there are always at least two elements in each MPLAB X project:

- *Makefile*, this is exactly what it claims to be, an (automatically generated) make file that will be used by the GNU *make* tool to build your project

- *nbproject* folder, this is where MPLAB X stores the configuration of your project, including the list of source file names to be compiled, linker scripts, your personal preferences, debugging tools selections and so on

NOTE
The name of this folder is revealing the origin of MPLAB X. It has never been a mystery that MPLAB X is based on the NetBeans IDE project.

Needless to say, you don't want to mess with the contents of these two, as both are generated and maintained automatically by MPLAB X. But what you might find useful is that, if you stick to some basic guidelines that I will highlight in the following chapters, you can make sure that the entire folder contents are *position independent*. This means that you will be able to move (or copy) the entire project folder to a different path on your hard drive or on a different machine and the project will remain intact and fully functional.

Logical Folders

So what is the relationship between the *logical folders* (see Figure 1.6) and the contents of the actual physical folder (Figure 1.7) we just revealed?
It is actually much more loose than you think. The *logical folders* are simply lists of file names. The location of those files is independent of the actual position of the project folder we just created. This is a powerful feature that we will exploit to our benefit to define "shared" resources, for example. At the same time, it is a potential source of confusion and eventually bugs if not understood and managed.

Hopefully the rest of this chapter, in its extreme (coding) simplicity will help clarify some of these concepts.

The most important logical folder is the one named: *Source Files*. ALL and ONLY files listed in here will be compiled and linked into our applications regardless of their location in the file system.
On the contrary, files listed in the *Header Files* logical folder, are there merely for our convenience. You can leave this folder/list empty if you want and your project will compile just fine! You can try this on any of the projects presented in this and the following chapters of the book.
Obviously, it pays to be disciplined and list in here the most important header files in a project. I recommend you do so to help document and maintain the project.

Similarly the *Important Files* folder contains just a link to the makefile. It is there for your reference. Feel free to inspect its contents with the built in editor (double click on the file name), but don't try to modify it by hand. Adding other files to this folder won't make a difference to the project build process.

The *Linker Files* folder will be important for those of you planning on using a custom linker script as is the case with the Microchip HID bootloader for example. The Mikromedia bootloader by contrast does not require a custom linker script and the same is true for all other programmers/debuggers, so this folder will remain empty and unused in most all circumstances.

The *Library Files* and *Loadable Files* logical folders will not be used as well in the following chapters as they refer to special (as in rare) use cases.

New File Checklist

Time to create the first source file. The *New File Wizard* (activated by the **CTRL-N** command, ⌘-N for MAC users) will assist you, offering a staggering number of options to help you quickly create the file from a list of templates.

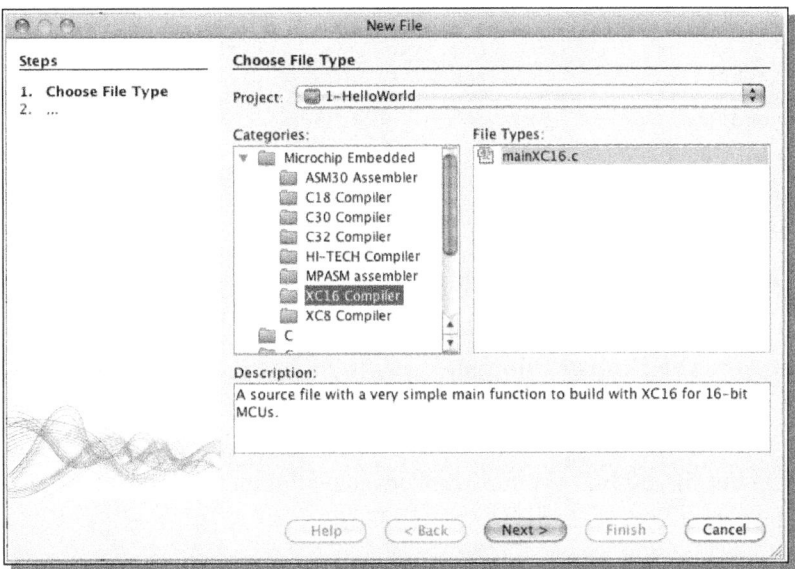

Figure 1.8 – New File Wizard

The New File Wizard is composed essentially of two dialog boxes and requires the following steps:

1. *File Type selection*: In the Categories pane, select **Microchip Embedded**
2. This will expand into a list of sub-categories, select **XC16 Compiler**
3. In the right pane titled *File Types*, select the **mainxc16.c** type
4. Click **Next**
5. The new *Name and Location* Dialog box will appear, here most fields will be already pre-filled with the default settings of your project (folder), you will have only to assign a proper name to the new file: type **main.c**
6. Click **Finish**

MPLAB X will create the new *main.c* file and will fill it with a basic template that is composed of the following few lines of code:

```
/*
 * File:   main.c
 * Author: (your name here)
 *
 * Created on (date and time here)
 */

#include "xc.h"

int main( void )
{
    return 0;
}
```
Listing 1.1 – mainXC16.c template

The alternative, perhaps more common case of use of this Wizard, is to create an empty file and then type your way through it. In this case, in the first dialog box you will choose the **Other** category and the **Empty File** type.

Note that the wizard not only creates the file and populates it with a template as required (see Listing 1.1) but, if you had the *Source Files* logical folder open, it does also add the newly created file to it.

Hello Embedded World

Every respectable programming book must contain a "Hello World" example. In the embedded world this is not necessarily done using text on a screen/terminal, although we will do that too very soon, but often simply giving an indication of activity by means of an LED blinking.

In the specific case of the Mikromedia board, this simple act might actually prove to be trickier than you'd expect. The fact is that none of the two LEDs present on the backside of the board, opposite to the LCD display, is directly under our control. The first one is purely indicating presence of power supply (3V). The other one is indicating the charging of a (lithium) battery, should it be connected and indeed charging.

Only if we look more carefully we can identify a third option: the LCD display backlight! This realization forces us to crack open for the first time the PIC24 Mikromedia documentation and to peek at the schematic to familiarize ourselves with the hardware at hand.

There is actually not just one, but a whole string of LEDs inside the (MultiInno MIO283) TFT display module and it is controlled by a simple configuration of bipolar transistors (see detail in Figure 1.9) which keeps the backlight of the display on by default at power up, but can be turned off using the LCD-BLED signal.

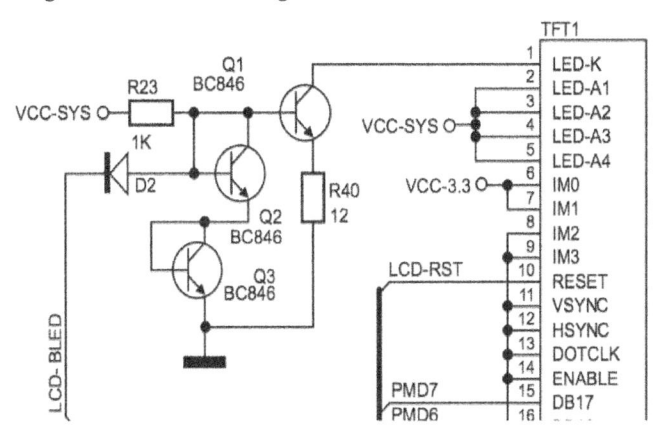

Figure 1.9 – Mikromedia board schematic, backlight control detail

If we follow the LCD-BLED line back to its source, we will notice that it is connected to pin 77 of the PIC24, corresponding to pin 2 of the general purpose I/O port D (aka RD2). Defining a couple of simple macros will help keep our code readable:

```
#define LED_BLED                _RD2
#define ConfigureBacklight()    _TRISD2 = 0
```

As a refresher, I will mention here that the notation _RD2 is truly a short for the more verbose PORTDbits.RD2 expression that selects the specific pin 2 of the I/O PORTD register of the PIC24.

Similarly _TRISD2 is a short version of the TRISDbits.TRISD2 control bit that defines the direction of the I/O pin. A '0' is used to define the pin as an output and a '1' defines the pin as an input. An easy way to remember this is that '0' resembles the letter 'O' for output and a '1' resembles the letter 'I' for input.

Using Timer1 to provide a precise delay of 250ms, we can then assemble our (first) version of "Hello World" for the Mikromedia board as follows:

```c
int main( void )
{
    // 1. init I/O and timer
    ConfigureBacklight();       // configure I/O as output
    T1CON = 0x8030;             // internal clock/2 /256

    // 2. main loop
    while (1)
    {
        TMR1 = 0;               // delay
        while ( TMR1< DELAY);

        LED_BLED = 1 - LED_BLED;// toggle LED_BLED output
    } // main loop

} // main
```
Listing 1.2 – Blinking the LCD display backlight

Once more, as a refresher, I will point out that `T1CON` is the main control register of Timer1, a basic 16-bit timer. Assigning it the value `0x8030`, as you can verify with a quick glance at the PIC24 datasheet, has the effect of enabling the timer and assigning it the main peripheral clock source with a prescaler of 1:256.

So we are left with only one simple problem, defining the value of `DELAY` to obtain the correct temporization.

If we assume that the PIC24 clock is set for the maximum speed of 32MHz, all the peripherals will receive a clock that is exactly ½ of that value, i.e. 16MHz. Applying the 1:256 divider in front of Timer1, we are actually ensuring that the timer will receive 62,500 ticks every second (16,000,000/256). So ¼ of a second will take exactly 15,625 counts. By defining `DELAY` as a macro, we can make this definition self explanatory as follows:

```c
#define DELAY     16000000UL / 256 / 4
```

From MPLAB X main menu, select **Run>Clean and Build Project** and if all goes well (you did not make any typing error) you should get a successful build and the following message should appear at the bottom of the *Output Window*:

```
BUILD SUCCESSFUL (total time: 406ms)
Loading code from /.. production/1-HelloWorld.X.production.hex...
Loading symbols from /..production/1-HelloWorld.X.production.elf...
Loading completed
```

Mikromedia Bootloader Notes

If you chose to use the bootloader to load the applications onto the Mikromedia board, you will have to follow a few additional steps before you'll be able to test the application.
If you have chosen any of the other options involving a Microchip or MikroElektronika in-circuit programmer, you can skip directly to the next section.

You must know that the native Mikromedia bootloader is in its turn a small application that was written to occupy the smallest amount possible of flash memory space but, nonetheless, is taking up *some* space. It is possible that an application might attempt to use the exact same memory space already used by the bootloader, a conflict that the bootloader would avoid by simply omitting any overlapping portion of the application. Fortunately, since the MPLAB XC16 linker starts filling the flash memory starting from the lowest addresses and the Mikromedia bootloader has been designed to occupy the highest possible memory address space, only the largest applications (256K byte large) will run into such possible conflicts and, for the most part, the exercises offered in this book will work without need for additional care.

Microchip HID Bootloader Notes

This is not the case for the Microchip HID bootloader which occupies the lowest section of the program memory. When using it, you will need to plan on adding a customized *linker script* to your projects. This is described in detail in the Microchip HID bootloader application notes and is available for download as part of the same package.
You can include the custom linker script in your project by adding it to the **Linker Files** logical folder in the **Project** window. Here is a simple step-by-step procedure:

1. Using a file manager, **copy** the Microchip HID bootloader linker script **.gld** file into your current project directory (**1-HelloWorld.X** in this example)
2. In MPLAB X Project window, select the **Linker Files** logical folder
3. Right click on it to open a context menu, select **Add Existing Items...**
4. In the file selection dialog box, make sure the **Store Path as** option is set to **Relative**
5. Select the **.gld** file

Finding the Executable File
Build the project with **Run>Clean and Build Project.**
The output binary *(.hex)* file will be found inside the MPLAB X project folder nested deeply down the following relative path:
dist / default / production / 1-HelloWorld.X.production.hex

Mikromedia Bootloader Programming Steps
Once you have launched the MikroElektronika USB HID Bootloader GUI (see Figure 1.10):
1. **Reset** the board and (within 5 seconds)...
2. Click on the **Connect** button to establish a connection.
3. Choose the .hex file by pressing the **Browse for Hex** button
4. Press the **Begin Uploading** button to transfer the code to board.

PICKit3, ICD3 or REAL ICE Programming Steps
If you have selected a PICKit3 (or superior) in circuit programmer, you can proceed to:
1. Use the MPLAB X **Make and Program** command, or
2. From the main menu you can select **Run>Run Project**, or use the keyboard **F6** shortcut (MAC users will have to press **Fn+F6**)

With a single mouse click, you will build the project and proceed to program the output binary file directly to the target board!

mikroProg Programming Steps
If you have selected the *mikroProg* tool, you will have to:
1. Build the project in MPLAB X first (**Run> Clean and Build**) and then
2. From the mikroProg GUI, **load** manually the output binary (.hex) file produced by MPLAB X

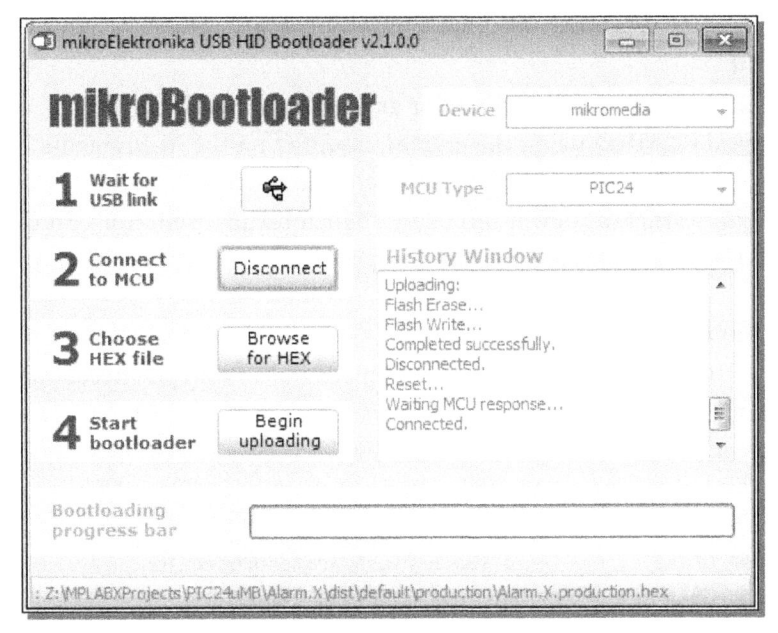

Figure 1.10 – mikroBootloader GUI

Testing Hello World

If you completed successfully one of the above programming sequences, you will be pleased to see the entire surface of the LCD color display blinking a bright white. With a precision instrument you could now check the blinking period, to verify that it is indeed 2Hz since the display will be on for 250ms, and off for 250ms for a total of two cycles a second.

Except that you won't need a precision instrument ... the timing will be off ... and by a large amount, something you will be able to tell immediately. In fact the frequency will be off by as much as two orders of magnitude! What's going on?

Not to worry, I enjoy playing this kind of tricks to my readers in the early chapters; they are meant to keep you alert and curious!

> **SPOILER ALERT**
>
> If you have not figured it out yet and you enjoy the challenge, take your time and re-check the code we entered so far.
>
> I will reveal the solution on the next page!

Mystery Solved

In the timing calculations we have taken for granted that the PIC24 would be running at its maximum speed (32MHz) using the crystal mounted on the Mikromedia board. Unfortunately this is not the case unless we *configure* the device to do so. By default, all PIC microcontrollers start (at power on) using their internal oscillators. In the case of our PIC24FJ256GB110 this is an 8MHz oscillator, exactly 4 times slower than we assumed.

Configuring the PIC24

Configuring the PIC24 with the proper oscillator source and all the necessary additional settings to obtain a 32MHz clock is simple enough if we use the *_CONFIG1* and *_CONFIG2* macros as follows:

```
_CONFIG1( JTAGEN_OFF        // disable JTAG interface
      & GCP_OFF             // disable general code protection
      & GWRP_OFF            // disable flash write protection
      & ICS_PGx2            // ICSP interface (2=default)
      & FWDTEN_OFF)         // disable watchdog timer

_CONFIG2( PLL_96MHZ_ON      // enable USB PLL module
      & PLLDIV_DIV2         // 8MHz/2 = 4Mhz input to USB PLL
      & IESO_OFF            // two speed start up disabled
      & FCKSM_CSDCMD        // disable clock-switching/monitor
      & FNOSC_PRIPLL        // primary oscillator: enable PLL
      & POSCMOD_XT)         // primary oscillator: XT mode
```
Listing 1.3 – PIC24 configuration

In detail, with the first configuration word we disable the code protection mechanisms and the watchdog – always a good idea when developing and debugging an application. We also disable the JTAG port that is not used by Microchip and MikroElektronika in circuit programming tools, gaining back four additional I/Os.

With the second configuration word, we take care of selecting the external 8MHz crystal source, we divide it by two to obtain 4MHz, and then enable the PLL to produce a 96MHz output. Finally we divide it by 3 to generate the desired 32MHz.

If it sounds a bit contorted, it is!

But it works and it is very flexible allowing us to handle a full speed USB connection (more on this in Chapter 8) and, at the same time, use a scalable source for the PIC24 core and its peripherals.

Rebuild the project and run it on the Mikromedia board to enjoy a true 2Hz big *blinky*!

Summary

In this first chapter we have spent a few extra pages to introduce the fundamental tools we will be using in the rest of the book and to install and configure them properly.

We have also discussed the choice of programming options available with their respective pros and cons.

Finally we have created a first Hello World project (or perhaps we should call it just Big Blinky) as an excuse to start familiarizing ourselves with the structure of an MPLAB X project, its directories and, in the process, brush off some rust in our PIC programming skills.

Admittedly this first example is not representative of the kind of projects we will be confronting in the coming chapters, but we had to start somewhere and I could only cram so much material in these first few pages. Hopefully the low level approach I used here (to avoid introducing any *libraries* too early on) did not discourage or scare you just yet.

Take a break now and continue reading into the next chapter tomorrow as you give yourself some time to digest this material properly.

Tips & Tricks

If you want to create your own *file template* or customize one of the existing ones, select **Tools** from the main menu and then choose **Templates**. This will give you access to the *Template Manager* where you will be able to add your own Categories and Files types.

Two additional techniques can be used to speed up typing:

- *Code Templates,* can be used in the editor to accelerate typing of common commands, and/or frequently used symbols. For example, type *"sw"* followed by CTRL+\ (⌘+\ for Mac users) to see the text expanded to *"switch ("*.
  Code templates are defined in the editor's preferences: open the **MPLAB Preferences** menu, select the **Editor icon** at the top of the dialog window and then select the **code templates** tab to edit the predefined set and/or add new ones.

- *Macros,* allow the recording of any sequence of keystrokes to help you with repetitive typing activities but can also be used to capture an entire template for a function header or a file template composed of any number of lines of text.
  To record a macro, select from the main menu **Edit**, then **Start Macro Recording.** Proceed typing as you would normally. When finished, **Edit>Stop Macro Recording** and assign a name (and optionally a shortcut) to the macro.

Suggested Reading

Lucio Di Jasio, "**Programming 16-bit microcontrollers in C. Flying the PIC24**", Second edition, Newnes Elsevier

- Chapters 1-3 covers project creation and I/O and timers initialization
- Chapter 5 covers usage of the Real Time Clock module and Interrupts
- Chapter 6 covers linker scripts and memory allocation

Exercises

1. Use the Real Time Clock peripheral module and the 32 KHz crystal mounted on the Mikromedia board to provide accurate timing and put the device in *low power mode* in between blinking intervals to save power.

   HINT

   > Look for help in the peripheral library (*rtcc.h*) or refer to the "Flying PIC24" book (see suggested reading).
   >
   > Also the *Sleep()* macro will allow you to put the microcontroller in low power (a.k.a. sleep) mode.

Solution

```c
/*
 * File:    main.c
 *
 * Project: 1-Solution.X
 *
 */
#include "xc.h"
#include <rtcc.h>

_CONFIG1( JTAGEN_OFF          // disable JTAG interface
        & GCP_OFF             // disable general code protection
        & GWRP_OFF            // disable flash write protection
        & ICS_PGx2            // ICSP interface (2=default)
        & FWDTEN_OFF)         // disable watchdog timer

// no need to enable the external 8MHz crystal osc and PLL
// to save power, use the internal oscillator (default) and rely on the RTCC
// for accurate timing

//_CONFIG2( PLL_96MHZ_ON       // enable USB PLL module
//        & PLLDIV_DIV2        // 8MHz/2 = 4Mhz input to USB PLL
//        & IESO_OFF           // two speed start up disabled
//        & FCKSM_CSDCMD       // disable clock-switching/monitor
//        & FNOSC_PRIPLL       // primary oscillator: enable PLL
//        & POSCMOD_XT)        // primary oscillator: XT mode

#define LED_BLED              _RD2
#define ConfigureBacklight()  _TRISD2 = 0

#define __RTCC_ISR    __attribute__((interrupt, shadow, no_auto_psv))

void __RTCC_ISR _RTCCInterrupt( void)
{
    LED_BLED = 1 - LED_BLED;   // toggle LED_BLED output
    _RTCIF = 0;
}

int main( void )
{
    // 1. init I/O and timer
    ConfigureBacklight();      // configure I/O as output
    RtccInitClock();           // init 32kHz oscillator

    __builtin_write_RTCWEN();  // unlock RTCC registers
    mRtccOn();                 // enables rtcc, also clears alarm cfg
    ALCFGRPT = 0xc000;         // enable alarm, chime, 1/2 second period
    _RTCIE = 1;                // enable alarm interrupts

    // 2. main loop
    while( 1)
    {
        Sleep();               // go to sleep

    } // main loop

} // main
```
Listing 1.4 – Solution, main.c

Chapter 2

Hello MLA

In this chapter we will start exploring the *graphic library*, one of the most popular components of the Microchip Libraries for Applications collection (from now on, simply MLA). This will help us get a first look at some of the essential mechanisms used throughout the MLA to allow it to accommodate for a large number of hardware modules, microcontroller models and the disparate assortment of interfaces used to communicate between them. This will be our first encounter with the *Hardware Profile* (that we will start customizing specifically for the PIC24 Mikromedia board) followed by the *Graphics Configuration* file, the first of a long series of configuration files we will encounter through the rest of this book.

We will also pay special attention to the proper positioning of the MLA files (relative to the project folders) to help us maintain a simple and consistent path to reach every library module we will need in present and future projects. Special considerations will also be made to ensure ease of maintenance, traceability and proper project (self) documentation.

Preparation

For this lesson, I will assume that you have followed satisfactorily the instructions in the previous chapter to:

1. Install MPLAB® X and the XC16 compiler
2. Create a working directory (*Mikromedia*)
3. Have bookmarks (links) to the key pieces of documentation
4. Have your programming/debugging tool of choice ready and connected to the Mikromedia board.

Getting the "right" MLA

Without further ado, we will proceed to download from Microchip web site the appropriate release of the Microchip Libraries for Applications.

> **ONLINE RESOURCES**
>
> This link will take you directly to the main download page:
> http://microchip.com/mla

Current MLA vs. Legacy MLA

Since December of 2013 a new and vastly incompatible version of the MLA has been published. The changes introduced are meant to help the MLA converge gradually toward the *Harmony* project (dear to the PIC32 users). Unfortunately beside the API modifications required, some of the tools, such as the GDD used in Chapter 7, are not yet ready to support it.

Eventually, as the new or now "*Current MLA*" improves and stabilizes, I will post updated source codes and porting notes using the book web site and blog to provide guidance and support as appropriate.

For the time being, I encourage the reader to still refer to what is now considered the "**Legacy MLA**" (a name that refers to the June 2013 release), a stable platform on which this book many examples and projects are based.

Library	Current Version	PIC16F (8-bit)	PIC18F (8-bit)	16-bit	32-bit
USB Framework	2.9h	x	x	x	x
Graphics Library	3.06.02			x	x
Memory Disk Drive (MDD)	1.4.2		x	x	x
TCP/IP Stack	5.42.04		x	x	x
(TCP/IP Stack v6.00.02 beta available for download below)	v6.00.02 beta			x	x
mTouch Capacitive Touch Library	1.42.00	x	x	x	x
Smart Card Library	1.02.8		x	x	x
MiWi™ Development Environment	4.2.6		x	x	x
Accessory Framework for Android™	1.02.00			x	x

Figure 2.1 – The MLA June 2013 release snapshot

As you can see from Figure 2.1, there are eight libraries included in the MLA collection and all of them operate seamlessly with any 16-bit and/or 32-bit PIC microcontroller.
Six of them are applicable to 8-bit PIC18 microcontrollers as well, and two of them are compatible with even the smallest of the 8-bit PIC architectures – the PIC16 group. Even then, the limitations are mostly due to the lack of sufficient RAM memory to cover efficiently the application at hand.

Downloading the MLA

From the MLA web page make sure to select the **Legacy MLA** tab and select the package that is appropriate for your operating system of choice.
Windows users get an executable (.exe) and Mac users receive a Debian image (.dmg) file as expected. Both types of users can simply double click the file to get the installation process going.
Things are a bit less automatic for the Linux users who receive a *.run* file.

> **NOTE FOR NOVICE LINUX USERS**
>
> Special considerations are required to turn this file into an executable and to activate its self contained installer.
>
> This is accomplished with the following two lines of code:
>
> ```
> sudo chmod +x microchip-solutions-v2013-06-15-linux-installer.run
> ```
>
> Replace the *.run* filename with that of the actual version you just downloaded.
>
> ```
> sudo ./microchip-solutions-v2013-06-15-linux-installer.run
> ```
>
> The *sudo* command will require you to enter your administrator password.
>
> Also pay attention to the second command: the dot preceding the first slash is quite essential.

Installing the MLA

No matter which installer you will be using, you will be asked to accept (click-through) a standard license agreement and, eventually, choose a directory where the contents of the library will be uncompressed into.
The place where you download the library contents is not so important as long as it is consistent with your regular workflow. For example, I have downloaded the files into the default *Downloads* folder of my Mac/PC and later told the respective installers to place the complete libraries in a subdirectory of my *home* folder (~/*MLA*) where I keep accumulating

(archiving) copies of each new revision. (Windows users will replace the ~ with C: and use backslashes).

Managing MLA revisions

Let's discuss for a second the implications of dealing with a set of libraries that is essentially in a constant state of flux. This can be a wonderful prospective as you have the certainty that there is a large team of developers and users worldwide that are busy at all times reporting and fixing bugs, continuously improving the libraries to support new features and new PIC models.

On the other side, this can be quite a scary scenario. Inevitably, more than once over the past several years (I believe at least some of the components of these libraries are close to reaching the first decade since birth) the changes have been somewhat disruptive, causing loss of backward compatibility and considerable headaches to those users that had not been careful documenting which of their projects required which specific version of the library (and kept backups of all sources).

It is imperative therefore that you apply some diligence in handling the update process and documenting your use of the libraries. At every new release of the MLA bundle, any number of libraries contained can change, and while there is always a *readme* file for each library documenting such changes, it can be quite a daunting task to keep track of them at all times.

So my recommendation is that you follow these simple practical guidelines.

When starting a new project:

- Start with the **2013 June release** (also referred to as *1306* for short in the following), and currently listed as the *Legacy MLA*.

- Make sure to document the MLA revision (date) inside your main project banner, so that two years from now (but often only a few months later) you, your team and/or other developer that will inherit your work, won't have to waste precious time hunting for clues now that any attempt to (re)build a project fails with a million mysterious error messages.

In general for existing and legacy projects:

- Keep the original, related MLA bundle in one of your regular backup directories.

NOTE FOR MERCURIAL AND GIT USERS

> I personally do not recommend you to include the entire MLA in your project repositories. At several hundred megabytes per copy, it is a considerable waste of space and time, *unless* you are working on extensions and modifications directly into the heart of the library. More on this later.

- In case of need, use the online archive (available via the "Archive" tab in the MLA download page) to access older revisions of the MLA.

- Note that the archive page contains previous revisions of the MLA that go back only as far as August 2009. It is also not clear for how long these archives will remain accessible although, knowing Microchip practices, this is likely to be well into the next decade and likely beyond.

MLA Under the Hood

Let's take a look at the folder where the newly uncompressed contents of the MLA have been deposited.

In case of the revision 1306 the installer will have created a folder called:

microchip_solutions_v2013-06-15

Inside it, you will find three types of subdirectories:

- *Starter Kit* specific, which will contain complete sets of projects customized for the individual Microchip Starter Kits of the PIC18, PIC24F and PIC24H

- *Demos and Combo*, containing collections of demonstration projects of various complexity that can be applied to the *Explorer16* and its *PICTails* with a very large number of *processor headers* (PIMs). *Smart card* and *Touch demos* are part of this set of folders as well.

- *Graphics, USB and TCP/IP*, containing support tools, GUIs and other useful resources for the respective type of projects.

- *Board Support Package,* containing small modules that are specific to functionality that is unique to individual development boards (including the Mikromedia board)

And last but most important:

- The *Microchip* folder. This is where the *real library source files* reside!

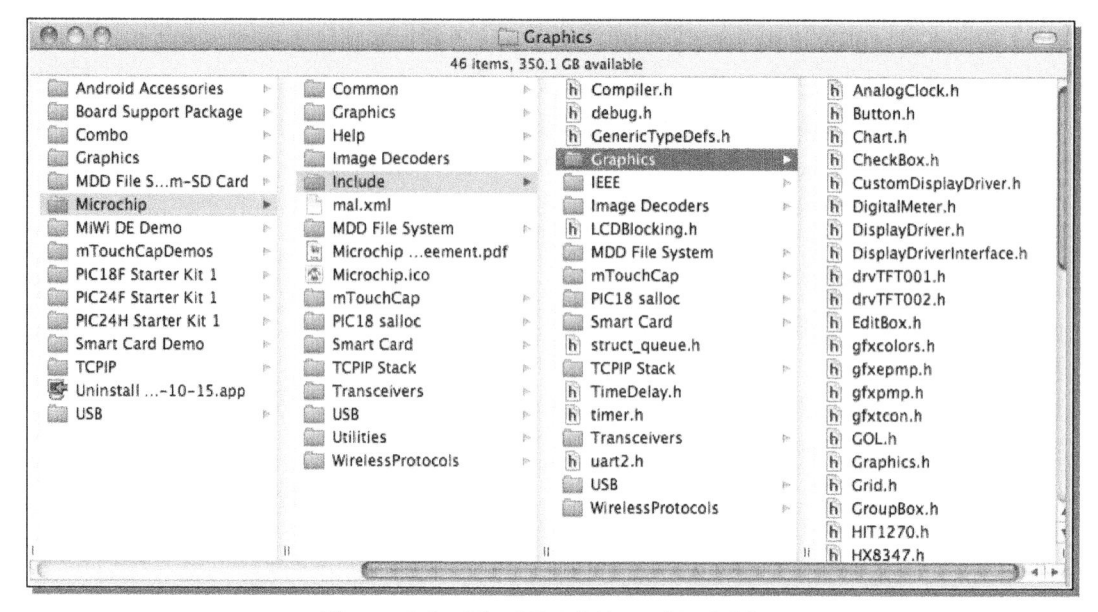

Figure 2.2 – The MLA/Microchip folder

Inside the Microchip Folder

Let's keep drilling, inside the *Microchip* folder (see Figure 2.2).
Here, explicitly enough, you will find a sub-folder for each major library: *Graphics, MDD File System, Smart Card, mTouchCap, TCPIP Stack, USB, Wireless Protocols*.
But you will also find two additional folders:

- *Help*, containing Windows help (*.chm*) and acrobat (*.pdf*) versions of each library documentation. These are automatically generated files that summarize the name and purpose of each library function and its parameters, as extracted directly from the headers/comments in the source files. You will find soon how the same information, can be obtained directly from the MPLAB X Editor, perhaps even more conveniently so while typing, using the *auto-complete* feature.

- *Include*, this is where all the header files for each library have been grouped. Inside it, you will find a new tree of folders that branches off with the same exact names of the (source) folders just one level above it.

In conclusion, all the MLA source files we will need to use in our projects are ultimately located inside the M*icrochip* folder and its subfolders. All the header files in particular, are located inside the *Microchip/Include* folder and its subfolders. Everything else around it is just documentation, examples and supplemental support tools.

We will use this knowledge to reduce the amount of duplication on our hard-drives while at the same time keeping the project (group of projects) tight and easy to maintain inside our working directory.

A Bit of Structure in the Working Directory

In the first chapter I defined the *working directory* as the main folder inside which all our PIC24 Mikromedia projects will be located. In my case, I used the path:
~/MPLABXProjects/Mikromedia but, as mentioned before, you are free to choose different names and locations as long as we understand each other.
Inside it, we already created our first project *1-HelloWorld.X* and, if you did your homework, the *1-Solution.X* project is there too.
At this point our working directory should look like in Listing 2.1:

```
Mikromedia
    1-HelloWorld.X
    1-Solution.X
```

Listing 2.1 – Working directory contents

It is time to introduce the MLA library in our workspace and we will do it by **copying** the **Microchip** folder and all its contents **from** the recently downloaded and installed MLA package **into** our working directory.
You can also **copy** the entire ***Board Support Package*** folder into our working directory, for brevity though, I suggest we rename it ***uMB,*** with the added advantage of avoiding spaces in folder names.
In truth Mikromedia projects will only need a few of the support modules contained in the original *Board Support Package* folder. If you prefer, you can leave the *uMB* folder initially empty as I will point out in the following chapters when and which module will need to be copied. We will also use the *uMB* folder to store default configuration files and to share newly created ones.
Our new working directory should now look like Listing 2.2:

```
Mikromedia
        1-HelloWorld.X
        1-Solution.X
        ...
        uMB
        Microchip
                Include
                Help
                Graphics
                TCPIP
                ...
                USB
```

Listing 2.2 – Working directory with Microchip and uMB folders added

I did not spell out the entire directory tree here for the sake of brevity, but I hope you got the picture.

Additionally, in order to avoid bloating unnecessarily the working directory, you might consider removing the *Help sub*-folder copy. In fact its presence inside the M*icrochip* (source) folder is questionable, and it is foreseeable how, in the future, the MLA developers might pull it up one level. Similarly, there are other scattered folders containing *resources* (graphic images and icons for example) that clearly do not belong at this level of the folder tree or deeper below, but can take up several additional megabytes of space and can be removed for all practical purposes.

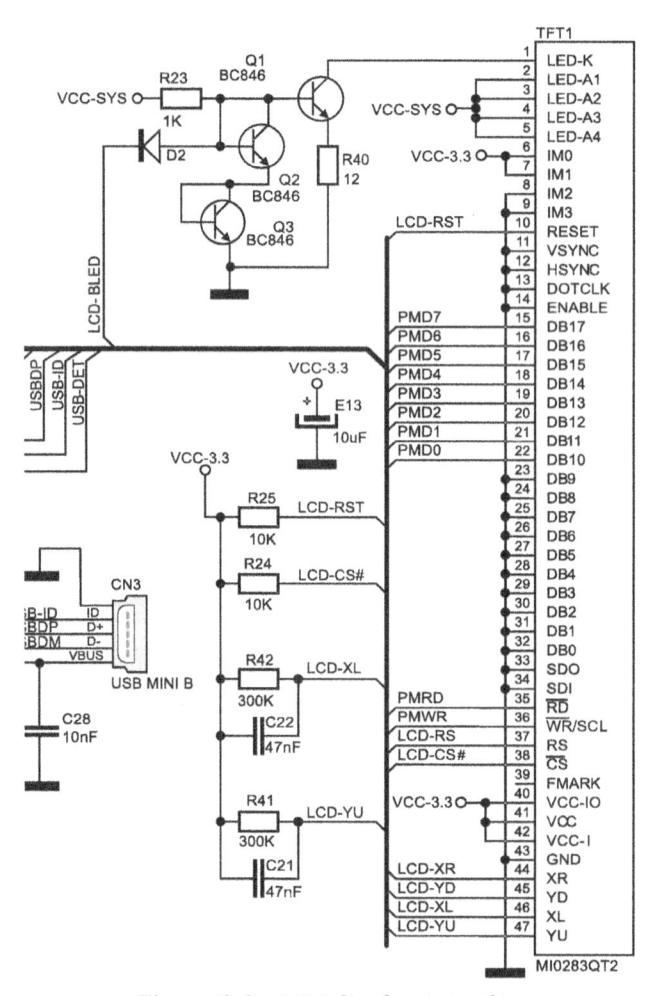

Figure 2.3 – LCD display interface

The Graphics Library

We'll begin now working on the first and, at least in the case of the PIC24 Mikromedia board, the most important MLA component: the Graphics library.
This will give us access to the LCD Color Display, certainly the most prominent feature of our board and an essential element of any embedded user interface application.

If we look at the PIC24 Mikromedia board schematic (see the detail in Figure 2.3), we can see that the LCD display module uses a MultiInno MIO283 display. This is part of the so-called *Mobile Display* generation. These are displays that integrate not only the traditional LCD (row and column) controllers but in the new technology known as Chip on Glass (COG), the LCD controller incorporates sufficient RAM memory to store an entire screen full of information – that is almost 256Kbytes of data. This relieves the main application processor, the PIC24 in our case, from some of the most mundane chores (refresh, windowing...) that would otherwise use up most of its bandwidth/performance.
The communication between the display controller and the application controller is handled via a bidirectional parallel bus (8 or 16-bit wide) that is at its lowest level not too dissimilar from what is used by the most inexpensive alphanumeric LCD controllers.

On the PIC24 Mikromedia board the display module is connected to the PIC24 via the 8-bit Parallel Master Port (*PMD0..7, PMRD, PMWR*) and a small group of digital I/Os assigned to control the *LCD-CS#, LCD-RS, LCD-RST* and four more analog inputs for the touch screen.
You will be pleased to know that the Graphics library, if properly *configured*, is perfectly capable of taking care of all these details for us, providing a clean and hardware independent set of primitives with which we will be immediately able to start drawing and printing text on the screen.
It would be nice to get a real "Hello World" message for once!

Configuring the MLA

There are two types of configuration details that we will be dealing with when working with the MLA:

- Hardware specific, such as pin and I/O assignments, communication interfaces (baud rates, polarity...) and timing. These configuration details are tightly connected to the PIC24 Mikromedia board design and are not going to change as we will progress through the book developing new projects. The MLA expects us to provide all such details (in the form of macro definitions) inside a single file: *Hardware Profile (or HardwareProfile.h)*

- Application/library specific, such as selecting a particular set of features available from any given library. These configuration details are relevant only on a per library basis and are likely to change from project to project. So we will have a Graphics Configuration file (or *GraphicsConfig.h*), a File System configuration file (or *FSConfig.h*), a USB Configuration file (*USBconfig.h*) and so on.

There are ample examples of such files scattered throughout the many demo projects included in the MLA bundle to draw inspiration from, but as often is the case, you CAN have too much of a good thing![2]

Over the years, in the herculean effort to support every possible combination of every existing PIC model with every possible hardware demo board, some of the configuration files offered in the MLA demo projects have turned into complex labyrinths that can scare and confuse the non-initiated and actually slow them down along the learning curve.
To avoid that, we will start slowly building our own hardware profile and configuration files piece by piece and, in the process, we will learn a lot more.

Creating a New Project

Borrowing from the previous chapter, we can reuse the *New Project Checklist* (based on the MPLAB X New Project Wizard) to start with a fresh new project inside our working directory. Let's call it: **2-HelloMLA**.

Inside it, let's create a new **main.c** file using the *New File Checklist* (based on the MPLAB X New File Wizard). The basic *XC16-main* template will do for now.
MPLAB X will create the file, populate it with the default *banner*, *include* directive and `main()` function and add it to the list of project source files in the corresponding *Source Files* logical folder as in Listing 2.3.

---

[2]Apparently the only known exception to this universal rule is: Nutella!

```
/*
 * Project: 2-HelloMLA
 *
 * File:    main.c
 */
#include "xc.h"

int main( void )
{
    return 0;
}
```
Listing 2.3 - HelloMLA, main.c

Creating the Hardware Profile

Let's launch the New File Wizard a second time. This time though, we will chose the **C++**[3] file category and choose the **Header file** type. Let's call it the *HardwareProfile.h*. This will give us a basic empty header file with the customary *#ifndef* wrapper to avoid multiple inclusions.

The MLA library source files will be "looking for" the Hardware Profile and will include it frequently in their headers. The order and number of times this happens would be difficult to control if it was not for the *#ifndef* wrapper.

```
/*
 * File:    HardwareProfile.h
 *
 * Hardware platform: PIC24 Mikromedia board
 */

#ifndef    HARDWARE_PROFILE_H
#define    HARDWARE_PROFILE_H

< place your definitions here >

#endif        /* HARDWARE_PROFILE_H */
```
Listing 2.4 - HardwareProfile.h in progress

Inside it, we can start immediately adding a first section containing three macro clock definitions:

```
/*******************************************************************
 * PIC24 clock
 *******************************************************************/
#define GetSystemClock()          (32000000ull)
#define GetPeripheralClock()      (GetSystemClock()/2)
#define GetInstructionClock()     (GetSystemClock()/2)
```
Listing 2.5 - HardwareProfile.h clock section

---
[3] The choice of the C++ category is meant to avoid the unnecessary *_cplusplus* wrapper that un-intuitively enough is part of the MPLAB X default C header file template.

These three macros will be used in multiple libraries inside the MLA to provide basic temporization, typically in short blocking loops, during device initializations procedures.

In our applications we might not want to use blocking loops altogether and/or most probably we will rely on timers to provide us with (more) accurate timing just as we did in the first example of Chapter 1. But the MLA tries to play nice. The code in the library is written to minimize the use of resources whenever possible and therefore, otherwise deprecated, *cycle-counting* loops are employed widely inside various device initialization routines.

NOTE FOR PIC EXPERTS

> You might object that these definitions create the false assumption that the PIC microcontroller and its peripherals will be bound to operate always at the same speed. In certain applications, it might be convenient to change the clock frequency dynamically to optimize power consumption and performance in different *states* of the application. In such cases, report here the clock values used during the *initial* device set up, at device power up and during the application initialization. Feel free to change the clock sources and configuration later as needed during the application main loop execution.

Creating PICconfig.h

Hopefully you enjoyed the little challenge at the end of Chapter 1. The lesson learned applies here as well. It is important to ensure that we do configure the PIC24 for the exact clock frequency that we declare in the hardware profile. In case of a mismatch, the MLA will most likely fail to initialize the various library modules/peripherals and this will cause the application to fail or to behave erratically.

As we did in Chapter 1, we can decide to add the two _CONFIGx() directives at the top of the *main.c* file or, as you can already see a pattern developing, we might want to create a little configuration file of our own here.

Let's invoke the **New File Wizard** once more and create a header file that we will call: ***PICconfig.h***. Mind this header file will need to be included only once (!) in each project, likely from the top of the *main.c* file.

In the rare occasions when in future projects we will need to depart from this basic configuration, we can simply omit including it and revert to the _CONFIGx() directives in its place.

Hello MLA - 47

NOTE

> As the symbols used inside the `_CONFIGx()` directives, such as `JTAGEN_OFF`, are specific to the PIC model used in the project and defined in the standard PIC support header files, it is necessary to ensure that *xc.h* is included before using them.

```
/*
 * File: PICconfig.h
 *
 * PIC24 Mikromedia projects device configuration
 */

#include <xc.h>

// this is the default configuration for all book's projects
_CONFIG1( JTAGEN_OFF        // disable JTAG interface
        & GCP_OFF           // disable general code protection
        & GWRP_OFF          // disable flash write protection
        & ICS_PGx2          // ICSP interface (2=default)
        & FWDTEN_OFF)       // disable watchdog timer

_CONFIG2( PLL_96MHZ_ON      // enable USB PLL module
        & PLLDIV_DIV2       // 8MHz/2 = 4Mhz input to USB PLL
        & IESO_OFF          // two speed start up disabled
        & FCKSM_CSDCMD      // disable clock-switching/monitor
        & FNOSC_PRIPLL      // primary oscillator: enable PLL
        & POSCMOD_XT)       // primary oscillator: XT mode
```

Listing 2.6 – PICconfig.h

Graphics Library Driver and Primitives Checklist

During our MLA Under the Hood exploration we have drilled quickly down the file system tree to reach the *Graphics* folder (see Figure 2.2). In it we found a relatively large number of files and a few subfolders. It is time now to select among them the few that are going to be needed to drive our LCD display and provide us with a first set of graphic primitives. We will need to include these files in our project *Source Files* logical folder. Here is a quick checklist that we will be using repeatedly in the next few chapters to enable our graphics applications:

1. In the *Project* window, select the **Source Files** logical folder and double click it to expose a context menu.

2. Select **New Logical Folder** to create a sub-folder.

3. Select the newly created (logical) folder and right click to expose the context menu. This time select **Rename...** and assign it the name **MLA.**

4. Right click on it again to expose the context menu again and select **Add Existing Item...**, this will open a file selection dialog box as in Figure 2.4

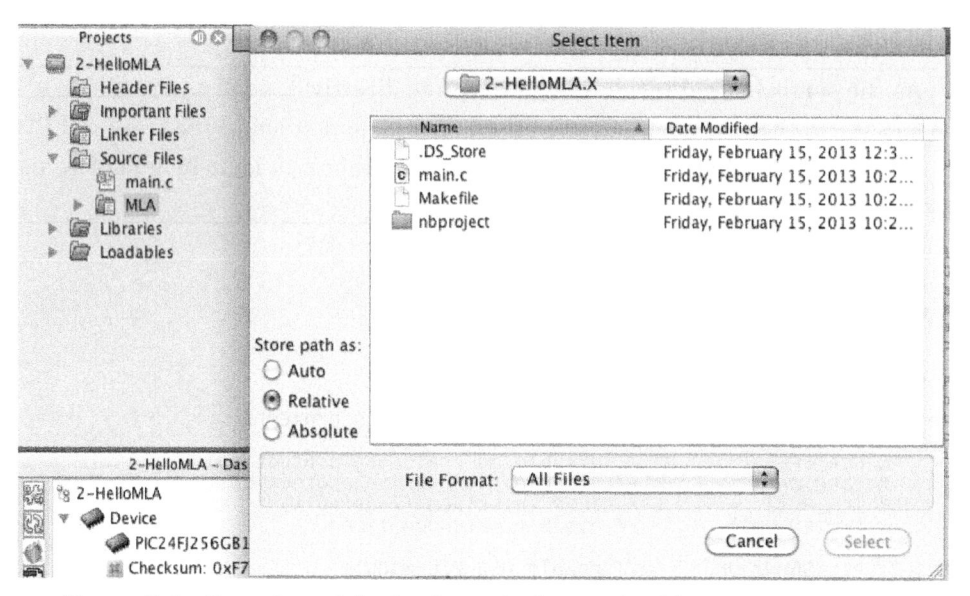

Figure 2.4 – Snapshot of the Project window and Add Item selection box

5. For a new project the file selection dialog box typically points to the current project folder. Ensure that the *'Store path as'* radio buttons are set on **Relative**

6. Then proceed to click on the folder name combo box (likely containing *2-HelloMLA.X*) and navigate one directory level up to the working directory (*Mikromedia*).

7. In there you will select the ***Microchip*** library

8. Inside it you will select the ***Graphics*** folder (we are drilling inside the MLA now)

Depending on the specific Mikromedia model (and hardware revision) used we will need to select the appropriate display driver. A display driver contains the initialization and low level communication routines required by the Graphics library to interface with the display controller chip.

9. Older models of the PIC24 Mikromedia board featuring a MIO283**QT2** display (using a Hynix 8347D controller) will need the **HX8347.c** driver module found in the ***Drivers*** folder.

   Year 2014 and newer models of the PIC24 Mikromedia board featuring a MIO283**QT9A** display (using an Ilitek 9341 controller) will need the **IL9341.c** driver module.

NOTE

> As the LCD display market follows the consumer market rapid cycles, LCD display modules get frequently updated and any product featuring them eventually will have to deal with multiple display controller options. Mikromedia boards are no exception to this rule. The MLA graphics library *Drivers* folder contains many compatible display controller drivers already but it is also very simple to add support for new models by adapting existing drivers based on the vendor datasheet. Such is the case of the new **IL9341** driver which can be downloaded from the book companion web site.

10. Select the **MLA** logical (sub)folder and right click on it to select the **Add Existing Item..** command once more

11. The file selection dialog box will be still pointing to the *Drivers* folder, so select the combo box at the top and step back up one level into the **Graphics** folder.

12. Here select the ***Primitive.c*** file. This file contains the most basic (generic) graphic primitives commands to initialize the display screen, clear its contents, plot on it, draw lines, circles etc. etc.

The checklist continues from here with one last item (for now):

13. Select the **MLA** logical (sub)folder again and right click on it to select the **Add Existing Item..** command for a last time.

14. The file selection dialog box will be still pointing to the *Graphics* folder, so select the combo box at the top and step back up one level into the ***Microchip*** folder.

15. Here select the **Common** folder and inside it choose the **TimeDelay.c** file. This is where the Graphics driver will get its cycle counting delay loops!

This looks like a lot of work, but don't despair. Eventually with repetition, you will find all this quite logical and with a little discipline you will be soon reusing a lot of this material.

Adding the Display section of the Hardware Profile

Although we have already included in the project a driver (source file) that is specific to the (Hynix 8347D or Ilitek 9341) controller used in the Mikromedia display, we still need to define a few parameters that are related to the connection between the controller and

PIC24, and also parameters that are specific to the glass panel (the same display controller can be used in a multitude of LCD displays of many pixel resolutions and shapes).
Here the key macro definitions grouped in a new section that we will add to the *HardwareProfile.h* file we just created:

```
/********************************************************************
 * Horizontal and vertical display resolution
 ********************************************************************/
#define DISP_HOR_RESOLUTION             240
#define DISP_VER_RESOLUTION             320

// display type MIO283QT9A - IL9341 controller
#define GFX_USE_DISPLAY_CONTROLLER_IL9341

// for older display type MIO283QT2 - HX8347D controller
// #define GFX_USE_DISPLAY_CONTROLLER_HX8347D

// interface - 8-bit PMP interface
#define USE_GFX_PMP
#define USE_8BIT_PMP
#define PMP_DATA_SETUP_TIME             (0)
#define PMP_DATA_WAIT_TIME              (0)
#define PMP_DATA_HOLD_TIME              (1)

/********************************************************************
 * Image orientation (can be 0, 90, 180, 270 degrees).
 ********************************************************************/
#define DISP_ORIENTATION                90

/********************************************************************
 * IOs for the Display Controller
 ********************************************************************/
// Definitions for reset pin
#define DisplayResetConfig()            _TRISC1 = 0
#define DisplayResetEnable()            _LATC1  = 0
#define DisplayResetDisable()           _LATC1  = 1

// Definitions for RS pin
#define DisplayCmdDataConfig()          _TRISB15 = 0
#define DisplaySetCommand()             _LATB15  = 0
#define DisplaySetData()                _LATB15  = 1

// Definitions for CS pin
#define DisplayConfig()                 _TRISF12 = 0
#define DisplayEnable()                 _LATF12  = 0
#define DisplayDisable()                _LATF12  = 1

// Definitions for Backlight control pin
#define DisplayBacklightConfig()        _TRISD2 = 0
#define DisplayBacklightOff()           _LATD2  = 0
#define DisplayBacklightOn()            _LATD2  = 1
```

Listing 2.7 – HardwareProfile – Display configuration section

As you can see in Listing 2.7, it is all pretty self-explanatory. We specify:

- Display resolution (240x320) and the orientation of the display we prefer (Landscape)

- We tell the Graphics library driver that we will be using the Parallel Master Port to pass 8 bits of data at a time to the display controller. Timing details are extracted from the display controller datasheet.
- Finally, we specify the I/Os that will be used to control the Reset, Chip select (CS), Register Select (RS) and the Backlight control lines.

Customizing the Graphics Configuration file

As you have seen, the Graphics library is composed of a relatively large number of modules. Some of them are necessary, such as the driver and the primitive modules, other will provide incremental functionality (buttons, windows...). But there is a second degree of configurability built *within* most of them. Using simple symbol definitions as switches, we can activate more or less advanced features. This is done in the ultimate effort to keep the total code size to an absolute minimum to allow even the smallest and most inexpensive microcontroller to benefit from the library.

Conveniently all such switches (symbol definitions) are meant to be grouped inside a single file called *GraphicsConfig.h*.

The Graphics library comes with several complete examples of a *GraphicsConfig.h* file in the demo projects. For convenience, we can start with any one of them and simply customize it for our purposes, commenting and uncommenting the symbols we need, rather then typing them all in from scratch.

As you can see from Listing 2.8, the entire configuration file fits inside a single sheet of paper and for what we can see now, most/all the *switches* are disabled (commented). There are only two symbols that are defined:

- *COLOR_DEPTH*, which assigned the value 16, as the Mikromedia display is capable of such color resolution (the actual maximum resolution is 18 bit per pixel but 16 is the highest value currently supported by the Graphics library)
- *USE_GOL*, which is not strictly necessary when using only the graphics primitives, but if you bear with me, it will shortly come in handy as it is connected to the inclusion of a default font.

IMPORTANT NOTE

> Make sure to save a copy of the modified *GraphicsConfig.h* file inside the project directory (*2-HelloMLA.X*), but also keep a copy inside the *uMB* sharing folder, renamed *GraphcisConfigTemplate.h*, for future use in every and each project we will develop over the next several chapters.

```c
/************************************************************************
 * Configuration Module for Microchip Graphics Library
 * This file contains compile time options for the Graphics Library.
 ************************************************************************/
#ifndef _GRAPHICSCONFIG_H
#define _GRAPHICSCONFIG_H

#define COLOR_DEPTH     16

/************************************************************************
 * Overview: Blocking and Non-Blocking configuration selection. To
 *    enable non-blocking configuration USE_NONBLOCKING_CONFIG must be
 *    defined. If this is not defined, blocking configuration is assumed.
 ************************************************************************/
//#define USE_NONBLOCKING_CONFIG // Comment this line to use blocking configuration

/************************************************************************
 * Overview: Keyboard control on some objects can be used by enabling
 *    the GOL Focus (USE_FOCUS)support.
 ************************************************************************/
//#define USE_FOCUS

/************************************************************************/
//#define USE_TOUCHSCREEN          // Enable touch screen support.
//#define USE_KEYBOARD             // Enable key board support.

/************************************************************************
 * Overview: To save program memory, unused Widgets or Objects can be
 *     removed at compile time.
 ************************************************************************/
#define USE_GOL                    // Enable Graphics Object Layer.
//#define USE_BUTTON               // Enable Button Object.
//#define USE_WINDOW               // Enable Window Object.
//#define USE_CHECKBOX             // Enable Checkbox Object.
//#define USE_RADIOBUTTON          // Enable Radio Button Object.
//#define USE_EDITBOX              // Enable Edit Box Object.
//#define USE_LISTBOX              // Enable List Box Object.
//#define USE_SLIDER               // Enable Slider or Scroll Bar Object.
//#define USE_PROGRESSBAR          // Enable Progress Bar Object.
//#define USE_STATICTEXT           // Enable Static Text Object.
//#define USE_PICTURE              // Enable Picture Object.
//#define USE_GROUPBOX             // Enable Group Box Object.
//#define USE_ROUNDDIAL            // Enable Dial Object.
//#define USE_METER                // Enable Meter Object.
//#define USE_CUSTOM               // Enable Custom Control Object
                                   //(an example to create customized Object).
/************************************************************************
 * Overview: To enable support for unicode fonts, USE_MULTIBYTECHAR
 *     must be defined. This changes XCHAR definition. See XCHAR for details.
 ************************************************************************/
//#define USE_MULTIBYTECHAR

/************************************************************************/
#define USE_FONT_FLASH             // Support for fonts located in internal flash
//#define USE_FONT_EXTERNAL        // Support for fonts located in external memory

/************************************************************************/
#define USE_BITMAP_FLASH           // Support for bitmaps located in internal flash
//#define USE_BITMAP_EXTERNAL      // Support for bitmaps located in external memory

#endif // _GRAPHICSCONFIG_H
```

Listing 2.8 – GraphicsConfig template

GREEN Light

Time to re-open the *main.c* file and get some graphics action!
First of all let's include the main Graphics library header file:

```
#include "Graphics/Graphics.h"
```

This must happen after the inclusion of the *xc.h* processor specific header and, for clarity, we will include it after the *PICconfig.h* file as well.

Next we can proceed to initialize the Graphic library by calling the *InitGraph()* function.

To visually verify that the operation succeeded, we will then proceed to select the color green using the `SetColor()` primitive (GREEN is defined as an RGB565 coded 16-bit value in the *graphics.h* header). This is followed by the `ClearDevice()` function call with the effect of painting the whole LCD display with the last color selected.
Finally, let's not forget to turn the backlight on, or all that green will not be able to shine through. Here is the whole of the *main.c* file.

```c
/*
 * File:    main.c
 *
 * Project: 2-HelloMLA
 */

#include "xc.h"
#include "PICconfig.h"
#include "Graphics/Graphics.h"

int main( void )
{
    // 1. Display initialization
    InitGraph();

    // 2. Application initialization
    SetColor( GREEN);
    ClearDevice();
    DisplayBacklightOn();

    // 3. main loop
    while( 1);

} // main
```
Listing 2.9 – main.c - green light

Final Checks

Warming up the engine, let's do a final round of checks before we launch our application. After saving the newly edited *main.c* file, let's ensure that the project directory now contains all the necessary elements. You can verify this using your File Manager (Finder) or using the *Files* window in MPLAB X (see Figure 2.5).

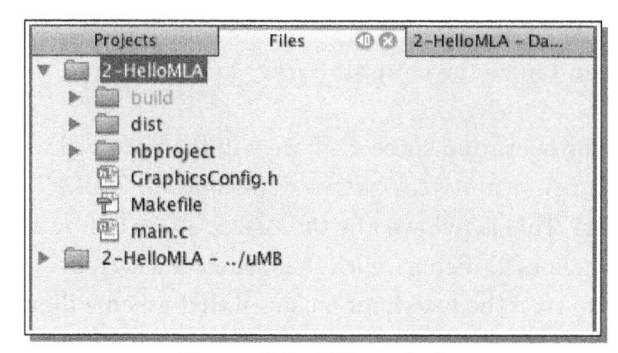

Figure 2.5 – Files window

NOTE

> Your project folder might not include a *build* and/or a *dist* sub-folder just yet. No worries, those will be created by MPLAB X shortly before launching the XC16 compiler and linker during the building process.

For documentation purposes, let's now add the *HardwareProfile.h* file and the local copy of the *GraphicsConfig.h* file to the *Header Files* logical folder in the Projects window:

- Select the **Header Files** logical folder and right click on it to expose the context menu.
- Then select **Add Existing Item..** to pick the two files from the project directory.

Remember, no matter what we put in the *Header Files* logical folder, it is only for our own reference. The XC16 compiler uses its own rules to find the include files based on the #*include* quotes used in your source file (" " or < >) and eventually the *global include path* settings.

Now see Figure 2.6 for an overview of the project as seen by MPLAB X:

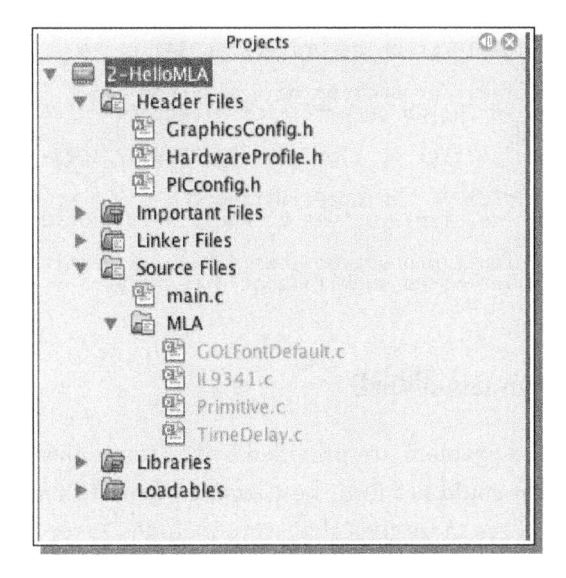

Figure 2.6 – Projects window

All looks right, let's build and launch the application!

As seen in the first chapter, the procedure will vary depending on whether you use a bootloader or an *in circuit debugger* such as the PICkit3, mikroPROG, ICD3 or REAL ICE.

- In the first case, use the **Run>Clean and Build Project** command first and then fetch the resulting binary (.hex) file from the *dist/default/production/* folder and bootload it to the board.
- In the latter case, simply use the **Run>Run Project** command to perform a project build followed by the automatic device programming phase.

Build Failure

And after all that careful preparation, I just did it again. I set you up for another failure!

Trust me, it is not cruelty. It is important that you see these error messages once, so in the future you will recognize them. Your MPLAB X *Output* window should contain a long and relatively messy listing resembling the (simplified) contents of Listing 2.10.

```
xc16-gcc   main.c  -o -c -mcpu=24FJ256GB110  -MMD ...
main.c:8:31: fatal error: Graphics/Graphics.h: No such file or directory
compilation terminated.

xc16-gcc   ../Microchip1210/Graphics/Drivers/IL9341.c  -o ... -c
-mcpu=24FJ256GB110 ...
make[2]: *** [build/default/production/main.o] Error 255
... fatal error: Compiler.h: No such file or directory
compilation terminated.
make[2]: *** [build/default/production/_ext/1410909387/IL9341.o] Error 255

xc16-gcc   ../Microchip1210/Graphics/Primitive.c  -o ... -c -mcpu=24FJ256GB110   ...
... fatal error: Graphics/DisplayDriver.h: No such file or directory
compilation terminated.
make[2]: *** [build/default/production/_ext/428596711/Primitive.o] Error 255
... fatal error: Compiler.h: No such file or directory

compilation terminated.
```

Listing 2.10 – Error window (simplified)

The clues to the root of the problem are provided by the text following the *fatal error* report. The XC16 compiler could not find the *Graphics/graphics.h* file we included in main. But it also failed to find three more files that were included respectively by the *primitive.c* module, *il9341.c* driver and the *TimeDelay.c* module.

I tried to warn you less than a page ago! The XC16 compiler does **not** get its include files paths from the *Header Files* logical folder in MPLAB X. For all practical purposes, the XC16 compiler is a completely separate entity from MPLAB X. The only rules that XC16 acknowledges for "finding" the include files are:

- If the #include <filename.h> (angled quotes) are used, the search is performed in the main XC16 library directory. From Chapter 1, you will remember our exploration inside the XC16 compiler installation directory where the standard C libraries and the peripheral libraries include files were found. In fact we got no complaints from the *xc.h* inclusion.

- If the #include "filename.h" (rounded quotes) are used, the search is performed in the *local* directory first (where the source file being compiled is located), and then eventually extended into the XC16 library directory. The files *Compiler.h* and *DisplayDriver.h* could not be found there. They are available inside the *Microchip/Include* folder inside the MLA instead.

There is only one way then to extend the reach of the XC16 compiler into the depths of the MLA and that is by defining a custom *include path* and feeding it to the XC16 compiler as part of its command line (-I command line option according to the XC16 compiler user guide). Fortunately MPLAB X can come to our rescue by allowing us to automate the process a little bit, but it won't do it without our input.

Setting the Include Path

The way we control how the XC16 compiler command line is formed by MPLAB X is through the *Project Properties* dialog box. You can open it by:

1. Selecting the current project in the *Projects* window and activating the context menu with a right click, then selecting the **Properties** option,
2. Or from the *Dashboard* window, by clicking on the little wrench tool shaped button,
3. Or from the *toolbar* clicking on the **default** configuration combo box,
4. Or... I am sure there are at least another five ways to do it!

Once the dialog box is open, take the following few steps:

1. Make sure to select the **xc16-gcc** category
2. **Check** the *Options Categories* selection box to select the **Preprocessing and Messages** collection
3. The **C Include Dirs** text field should appear as the top item on the scrolling list of options
4. **Click** on the button to the right of it to enter the options editor Dialog box (see Figure 2.7)
5. **Double click** on the first empty line, to activate a cursor and type:
   . (just a dot), then press **Enter**
6. **Double click** on the next empty line to activate a cursor and type:
   ../Microchip/Include (note the two "dots" in front of the first "slash")
7. **Double click** on the next empty line to activate a cursor and type:
   ../uMB (note the two "dots" in front of the first "slash")

   Alternatively, you could use the *Browse* button to navigate to the required folders instead of typing but make sure to select the ***Relative Path*** *(checkbox)* option to keep the project position independent.

8. Select the **OK** button to close the *C Include Dirs* editor
9. Select the **OK** button to close the *Project Properties* dialog box.

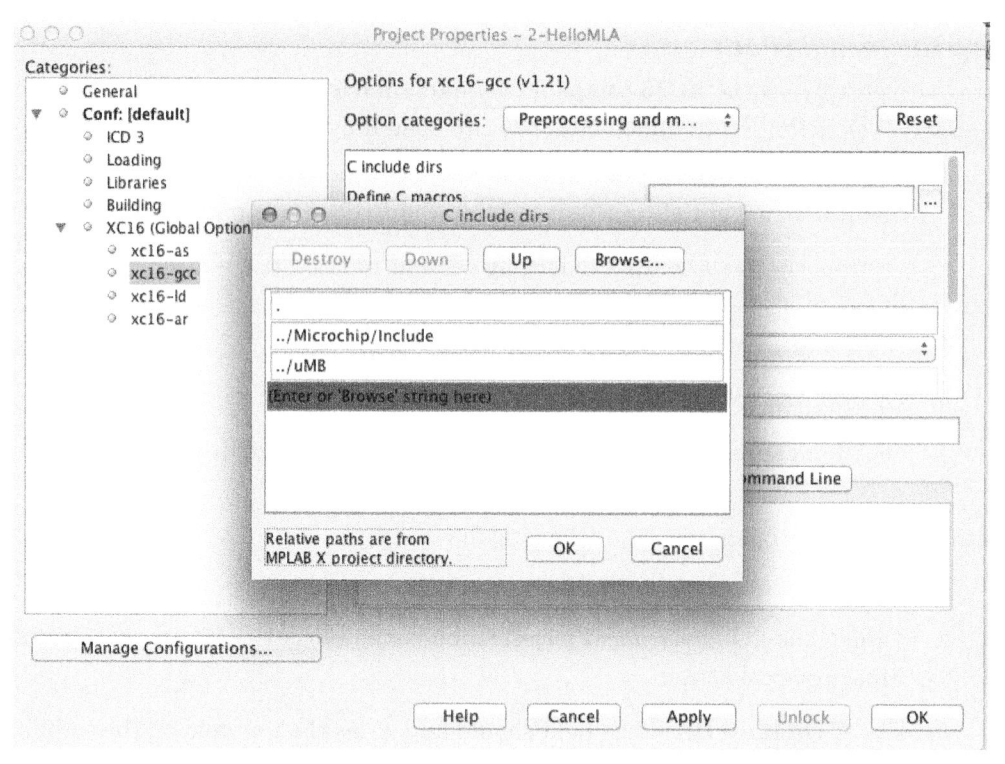

Figure 2.7 – Project Properties dialog – *C include dirs* **Editor**

The three paths we added have the purpose of creating a double link between the project and the Graphics library:

- **.** (dot), represents the current project directory. This allows the graphics library modules to reach into our directory and to grab the *GraphicsConfig.h* file we created.
- *../Microchip/include*, allows our *main.c* file, and other components of our application, to reach into the MLA and access the necessary header files (*Compiler.h*, ...).
- **../uMB**, gives the compiler access to the shared folder where *PICConfig.h* resides

These are all relative paths. They work within the current layout of the working directory, but they will continue to work in the future should you rename or move the entire working directory to a new location.

Second Attempt

With the include path set correctly and cross linking the project with the MLA modules, we can launch the make process once more and finally enjoy a, much deserved, successful build. Loading the application on the target, you will have the pleasure to admire the LCD display of the Mikromedia board shine a bright green light!

Playing with the Primitives

Onto new and amazing adventures. Let's see some of those promised graphics primitives in action, including: `MoveTo`, `LineTo`, `Rectangle()`, `Circle()`, `Bevel()`, and eventually `Bar()`, `FillCircle()`, `FillBevel()`.

Listing 2.11 contains a simple test sequence that will give you a taste of the graphics performance of the PIC24:

```
// 3. testing the primitives
    SetColor( BLACK);
    for( i=0; i<100; i+=10)
    {
        MoveTo( 10, 10);
        LineTo( 100, 10+i);
        Circle( 20+i, 220-i, 10+i);
    }

    SetColor( BRIGHTRED);
    Bar(120, 10, 170, 60);
    Rectangle( 120, 10, 200, 90);

    SetColor( BLUE);
    FillBevel( 220, 20, 250, 50, 10);
    Bevel( 220, 20, 310, 110, 10);

    SetColor( ORANGE);
    FillCircle( 270, 190, 40);
```

Listing 2.11 – Playing with the Primitives

Insert this code after the initialization (point 2.) but before the main loop in *main.c* and re-launch the project to enjoy!

Finally, Hello MLA!

The last *primitive* we are going to explore is the `OutText()` function.
This is what we ultimately need to send our "Hello" message to the world from the Mikromedia display.
Yes, you can rush to add *OutText("Hello MLA!");* to the project main file, but don't launch another build just yet.
There is one important resource that we are going to need first. We need to add a font!

A font can be a relatively large resource to add to an embedded application. Naturally, the size depends on the number of symbols it contains. Western fonts can use ASCII or basic subsets of it to keep space to a minimum but Asian fonts can be massive and potentially include thousands of symbols.
With the usual restrain and consideration for the cost of resources in the smallest embedded applications, MLA developers have provided means to choose selectively which symbols are going to be included on a *symbol-by-symbol* basis as necessary.
Also, while most modern PC applications can render and scale fonts *on the fly*, this is not a practical option in embedded control applications given the limited performance (and RAM space) available. So fonts need to be *pre-rendered* and then stored in a suitable format.
In the next chapter we will see which tools are available to perform such operation and how to package a required font resource for use in our applications.

For now, we will be satisfied with the default font offered by the *Graphic Object Layer* (or simply GOL from now on), a simple Graphic User Interface engine that is an integral part of the Graphics library.

> **NOTE**
>
> Now you understand the reason why, from the very beginning, I suggested that we leave uncommented the USE_GOL switch in the *GraphicsConfig.h* file.

Therefore *GOLFontDefault.c* is the last source file that we need to include in the *Source Files* folder of our project.
For the last time, select the **MLA** (sub)folder in the *Projects* window and right click to activate the context menu. Choose **Add Existing Item**.. and browse into the *Microchip/Graphics* folder where you will find the *GOLFontDefault.c* file.

> **NOTE**
>
> You can inspect this file, as it is a normal C source file. You will discover that it defines a structure called `GOLFontDefault` and initializes it with an array of integers representing each symbol (glyph) with a grid of pixels.

To use the font resource, we need to:

- Pass a pointer to it to the library with the `SetFont()` function
- With `MoveTo()`, we can set the top left corner of the box where the text string will be displayed
- With `OutText()` we can pass the string that needs to be rendered

Alternatively we can use `OutTextXY()` to perform the last two actions in one call.

Here is the last snippet of code that we have been working so hard to get at:

```
// 4. Hello World
SetFont( &GOLFontDefault);
SetColor( WHITE);
OutTextXY( 100, 100, "Hello MLA!");
```

Listing 2.12 – Hello MLA, last section

Tips & Tricks

Code Browsing

You can get a very detailed description of the primitives available from the Graphics Library Help (.pdf) file found in the Help folder (Chapter 6.3 of the manual) or, more simply, you can use MPLAB X to browse through the header file in a hyperlinked fashion.

- Starting from main.c, simply place the cursor on the *"Graphics/graphics.h"* part of the include statement and press **CTRL-B** (⌘-B for Mac users). This will immediately open the *Graphics.h* file in the MPLAB X editor.
- Select the first include statement that contains "primitive.h" (place the cursor on the file name) and press the **Browse** (CTRL-B or ⌘-B) command again. The editor will now show the contents of *primitive.h* and you will be able to scroll through it to inspect all the function definitions.

Now if you are curious to see how a particular function is implemented, simply place the cursor on it and right click to expose the context menu. Then select **Navigate** (a sub-menu) and inside it select **Go To Implementation**.

In general when inside a .h file you can quickly switch to the corresponding .c file (if MPLAB X knows the path to it) by using the **Go To Header/Source** button found on the far right of the editor toolbar.

Code Completion

As you familiarize yourself with the primitive functions names, don't spend too much time trying to memorize the exact sequence and type of their parameters. MPLAB X can provide that information on the fly as you type with the *code completion* feature.

Simply place the cursor in the editor window on a new line, then start typing *Fill* (for example) and press the **CTRL-** key combination. A small context menu will pop up offering you possible ways to complete the command with known function names selected from the header files recently included. Scroll to select the desired options among the available and press enter to transfer it into the source code. A small window will continue to hover over the cursor to indicate which parameter (and type) needs to be entered next and will follow you as you proceed through the completion of the function parameter list.

Summary

In this chapter we have started investigating the use of the first MLA library module: the *Graphics* library. We have started familiarizing with the way this module can be configured to the specific needs of our target board, the PIC24 Mikromedia, thanks to the abstraction provided by the *Hardware Profile* and the customization options in the *Graphics Configuration* file. We identified the key source files required to assemble a basic working project (*DelayTime.c, Primitive.c* and the graphic *driver*) and we learned how to set MPLAB X *C Include Dir* so that the XC16 compiler will be able to find all the required header files.

Finally, we explored the Primitive layer API of the Graphics library and eventually succeeded in sending our "Hello MLA" message to the world.

Suggested Reading

- **Graphics Library Help** (*Microchip/Graphics/Help/Graphics Library Help.pdf*),
    - Chapter 6.3 – Graphics Primitive Layer API
- Di Jasio, **"Programming 16-bit Microcontroller in C"**,
    - Chapter 10 - "Glass Bliss"
      Learn how to use the PMP peripheral to interface to an alphanumeric LCD display

Exercises

1- Centering text on the display

At this point a little geometrical problem would be ideal to help us truly understand how the text primitives work. Let's find out how to print a string of text on the display, nicely centered *horizontally*, and in a given color. In other words, the challenge is to implement a function whose prototype could look like this:

```
/**
 * @func   CenteredText
 *
 * @param y              vertical coordinate
 * @param color          color used for text
 * @param s              string to be centered
 */
void CenteredText( unsigned y, unsigned color, char* s)
```

Extra bonus points will be given if you also demonstrate how to use it (in `main()`) to center the line *vertically* on the screen.

> **HINT:**
> You will need to investigate the primitives:
>
> `GetTextHeight()` and `GetTextWidth()` as well as `GetMaxX()` and `GetMaxY()`.

2- Fading In

This exercise we will replace the *DisplayBacklightOn()* call in the initialization sequence with a call to a function that gradually increases the luminosity of the display from completely dark to maximum brightness over a programmable period (250ms).

```
void DisplayFadeIn( unsigned msTime);         // display fade in time (ms >100)
```

This will require you to look once more at the schematic of the Mikromedia board in order to refine the control of the display backlight.

> **HINT:**
> Configure the I/O controlling the display backlight as a PWM module output then increase/decrease gradually the duty cycle in small increments over the given period. The PPS.h peripheral library will help you to connect pin to function.

Solution 1

```c
/*
 * File:    main.c
 *
 *
 * Chapter 2: Solution
 *
 */
#include "xc.h"
#include "PICconfig.h"
#include "Graphics/Graphics.h"


/**
 * @func   CenteredText
 *
 * @param y              vertical coordinate
 * @param color          color used for text
 * @param s              string to be centered
 */
void CenteredText( unsigned y, unsigned color, char* s)
{
    unsigned margin;
    unsigned width = GetTextWidth( s, (void*) &GOLFontDefault);

    // check if too large to fit
    if ( width > GetMaxX())
        margin = 0;                              // use no margin, clip the right side
    else
    // compute margin to center
        margin = (GetMaxX() - width) / 2;    // split margin equally on both sides

    // set the color
    SetColor( color);

    // print string on screen
    OutTextXY( margin, y, s);

} // Centered Text

int main( void )
{
    unsigned height;

    // 1. init
    InitGraph();
    DisplayBacklightOn();
    SetFont( (void*) &GOLFontDefault);

    // 2. Centered Title
    height = GetTextHeight( (void*) &GOLFontDefault);
    CenteredText( GetMaxY()/2-height, BRIGHTRED, "Chapter 2: Solution");

    // main loop
    while( 1)
    {

    } // main loop
}
```

Listing 2.13 – Solution: CenteredText()

Solution 2

Add the following two include files:

```
#include "TimeDelay.h"
#include "PPS.h"
```

This is a simple 128 steps implementation:

```
/**
 * @func    DisplayFadeIn
 *
 * @desc    Gradually increases luminosity of the display backlight
 *
 * @param   msTime    (minimum 128ms, maximum 2000ms)
 */
void DisplayFadeIn( unsigned msTime)
{
    long w;

    // check time requested
    if ( msTime < 128)
        msTime = 128;                   // too fast, big steps
    if ( msTime > 2000) return;         // too slow, avoid overflows

    // connect pin RD2 to OC1 output
    PPSOutput( PPS_RP23, PPS_OC1);

    // configure OC1 block to generate a PWM signal
    OC1CON1bits.OCTSEL = 0x7;           // use peripheral clock (16Mhz)
    OC1CON1bits.OCM = 0x6;              // edge aligned PWM
    OC1CON2 = 0;                        // duty cycle
    OC1RS  =  0xffff;                   // set period  ~ 240Hz (16MHz/65.536)

    for( w=0; w<65536; w+=(65536/128))
    {   // 100 steps
        OC1R = (unsigned)w;             //
        DelayMs(msTime>>7);             // msTime/128
    }

    // Return display BL control to I/O
    PPSOutput( PPS_RP23, PPS_NULL);

    // Full brightness
    DisplayBacklightOn();

} // DisplayFadeIn
```
Listing 2.14 – Solution: DisplayFadeIn()

Chapter 3

Graphic Resources

In this chapter we will continue our exploration of the Graphics library, a fundamental components of the Microchip Libraries for Applications. We will review the management of resources such as fonts and bitmaps . We will also develop a simple terminal emulation module that, among other things, will be useful in future explorations as a quick debugging and prototyping tool.

Graphic Resources and Performance

In the past lesson we reviewed the primitives offered by the Graphics library to perform basic drawing commands up to and including the use of fonts.
As you have surely noticed, the graphics library does support proportionally spaced fonts. This is an efficient choice as it allows us to pack more text on the screen given the relatively low resolution available.
We must acknowledge that at approximately 100 pixels per inch, the little 3.2" QVGA display is far from *retina* resolution. But the combination of screen size and resolution is otherwise ideal for the class of microcontrollers we are currently using (16-bit x 16 MIPS and up to 32-bit x 80 MIPS). A larger screen or higher resolution would require much more work from the processor to update the screen contents and that would eventually result in a slower, less responsive user interface.

There are also considerations about RAM and FLASH memory requirements that need to be made. A QVGA display image is composed of 320x240 pixels, each containing 16-bit of color information giving a palette of 65K colors and adding up to approximately 150Kbytes of memory (320*240*2). For comparison, this is one order of magnitude larger than the amount of RAM available inside the PIC24GB1 series microcontrollers (16Kbytes) but is also comparable to the size of the entire flash memory available.

Since the display controller, present inside the display module itself, has all the RAM required to store the entire picture, the PIC24 needs only to provide updates addressing selectively small areas of the screen at a time.
The price for such convenience is paid though in terms of performance. The connection between the PIC24 and the display controller happens over the Parallel Master Port used as an 8-bit bus. This keeps the pin count down and reduces the complexity of the PCB, but forces the PIC24 to perform two consecutive transfers in order to update each single pixel on the screen. Since the chip select line (CS) of the display controller is managed via a

standard I/O, there are actually several more microcontroller clock cycles required to perform a single pixel update.

> **NOTE**
>
> In reality the MIO283 used in Mikromedia boards has a true *color resolution* of 18-bit per pixel, allowing 262,122 colors to be displayed. Accessing those additional two bits of color resolution though would require a 30% increase in pixel access time, quite a big performance penalty for a barely noticeable increase in color accuracy.

Alternative Microcontroller Choices

Several 16-bit microcontrollers of the PIC24DA series offer the possibility to control a color display more directly by using larger amounts of on chip RAM. But since the amount of internal RAM is still somewhat limited (as of this writing) to a max of 96K bytes, the maximum supported display size is correspondingly limited to QQVGA, that is a quarter of a quarter of a VGA screen, or 160x120 pixels. Alternatively a QVGA display can be limited to 256 unique colors, using an 8-bit memory array and a color palette, in which case the visual effect can still be quite compelling if the color palette is carefully assembled. PIC32 microcontrollers can instead count on up to 128Kbytes of internal RAM memory in the MX series and up to 512K bytes in the MZ series and can use DMA to drive the display with higher color resolution and size without any noticeable performance degradation.

> **NOTE**
>
> PIC32 *Mikromedia* boards as of this writing feature only the PIC32MX4 family limiting the internal RAM to 32Kbytes, but adopt a 16-bit wide PMP interface which more than doubles the communication bandwidth with the display.
>
> There are also PIC32 *Multimedia and PIC32 Mikromedia Plus* boards offered with a larger 4.2" display and with additional Ethernet and wireless communication interfaces featuring the PIC32MX7 family.
>
> All the examples and instructions provided in this book can be applied to those more powerful boards once the Hardware Profile is appropriately customized. Refer to the book companion web site for porting instructions and links to source code repositories.

Preparation

Continuing from where we left at the end of Chapter 2, we will create a new project and immediately populate it with the key components of the Graphics library. Here the simple steps:

1. Use the *New Project Wizard* to create a new project inside the working directory (*Mikromedia*), let's call it: "**3.1-Bitmaps.X**"
2. Add a new logical folder to the Source Files folder, called **MLA**
3. Add the following items to the folder:
    - ***IL9341.c** (or HX8347.c)*, the display controller driver depending on the display model
    - ***Primitive.c*** , the primitive layer of the graphics library
    - ***TimeDelay.c***, a few basic timing (blocking) functions used in the driver
    - ***GOLFontDefault.c***, a default font resource
4. Configure MPLAB® X *"C Include dirs"* to contain:
    - **.** (dot), the current project directory
    - **../Microchip/Include**, for our source files to reach inside the MLA
    - **../uMB**, for our source files to reach inside the shared folder
5. Create a new ***main.c*** file using the **New File** wizard and the embedded template (or your own customized version)

Bitmaps

While the Mikromedia display module is taking care of its own RAM buffer, the microcontroller is still required to produce any image that the application needs. Sometimes these can be generated using the primitive line drawing, filling and text functions we introduced in the previous chapter. But what if we need to display some more elaborate images and/or photographic images?

The simplest solution is to use bitmaps. That is, arrays of pixel color information representing rectangular areas, possibly as large as the entire screen.
By packing the information and sharing the processor (256K bytes) flash memory, we can make a number of such bitmaps immediately available to the application. Alternatively up to 1M byte of flash memory is available on the Mikromedia board as external serial

storage. But we will see that later, for simplicity, let's concentrate on the internal memory use first.

If we stick to the full 16-bit resolution of the display, a single bitmap the size of a fingerprint on the screen (64*64pixel) will require 8K bytes of flash space, ouch!

Depending on the content of the image though, we could use a color *palette* (effectively selecting only 256 unique colors per bitmap) to reduce the same array to approximately half that size, 4K bytes. A palette is nothing more than a table of a given number of entries each corresponding to a color. In the bitmap array, each pixel can now be represented with the index corresponding to the desired color (an 8-bit integer in this example) instead of the full 16-bit (or more) color information.

Similarly a reduction to a palette of 16 colors would reduce further the bitmap size to 2K bytes. Effectively each pixel would now require only a 4-bit index therefore the color information of two pixels can be packed in each byte of a bitmap array. Obviously this will be an acceptable compromise only for non-photographic images.

Reducing the bitmap size to 1K byte using only a monochromatic image (1-bit per pixel) represents the final limit but it's an option that is available only for a very restricted set of applications.

Primitive Bitmap Support

The Graphics library will help us transfer automatically any of the above four types of bitmaps to the screen if only we take care of formatting the image information appropriately.

The primitive function at our disposal is:

    PutImage(left, top, (void*) image, stretch);

- The *left* and *top* parameters are, as usual, referring to the x,y coordinates where the upper left corner of the bitmap will be placed.
- *Image* is a pointer to a bitmap resource, which is ultimately a structure containing the following information:
    3. Where the image is stored, *internal flash* or *external storage* device
    4. *Width and height* of the bitmap
    5. *Color Resolution*: 1, 4, 8, 16 bit per pixel

6. A *packed array* of pixel data (if in internal flash)
- The *stretch* parameter can assume two possible values:
    4. *IMAGE_NORMAL*, indicating a direct transfer of the image to screen
    5. *IMAGE_X2*, the bitmap is stretched horizontally and vertically so that each pixel in the bitmap is translated in a 2x2 square on the screen.

The Graphic Resource Converter

Preparing a bitmap resource by hand can be a seriously tedious task, so MLA developers have made available a specific tool: the *Graphic Resource Converter (GRC)* to facilitate the creation of a number of resources including bitmaps and fonts of course.

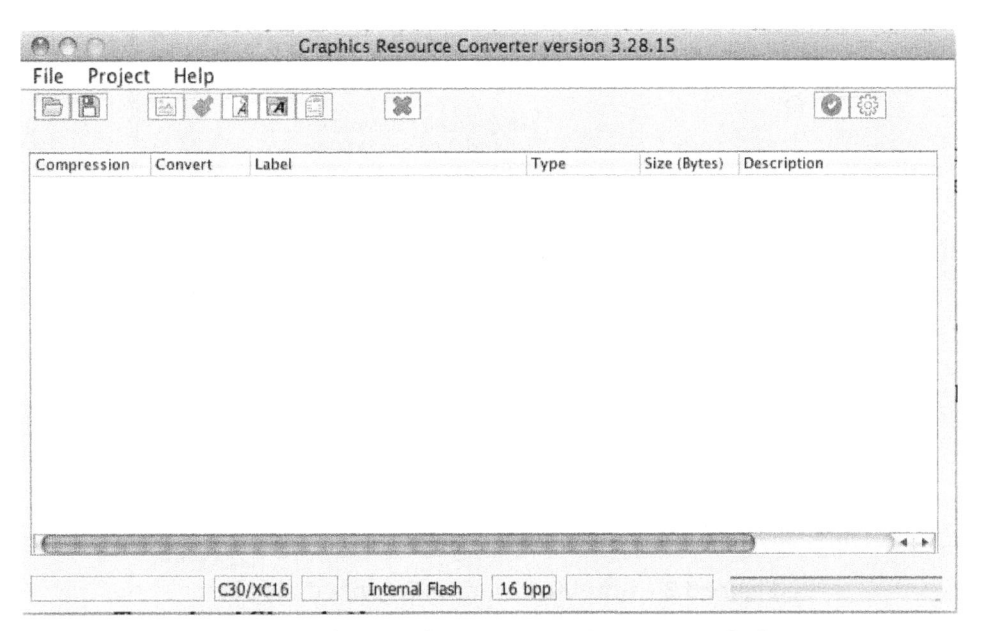

Figure 3.1: The Graphic Resource Converter window

The *Graphic Resource Converter* tool is found in the *bin* folder inside the *Microchip/Graphics* folder and is written in Java so to be available on all operating systems. Windows users will be able to invoke the **launch_grc.bat** file while Mac and Linux users will be able to simply double click on the **grc.jar** file to launch it.
The converter main window (see Figure 3.1) is almost entirely used up by a table, initially empty, that will list all the resources (images for now) to be converted for our project.
A small toolbar at the top and a thin status bar at the bottom complete the picture.

The first thing you will want to do is to configure the converter for use with the Mikromedia board display and the XC16 compiler. All the converter settings are available by selecting **Project/Settings** from the application menu, which is always available inside the Graphics Resource Converter window (instead of the top of the screen much to Mac users' surprise) or by clicking on the little gear button.

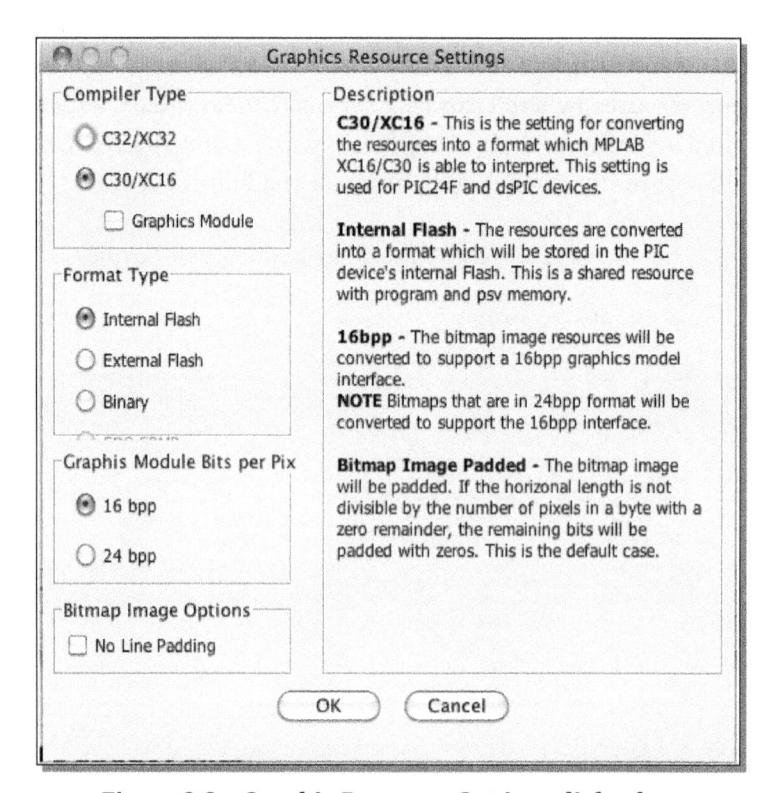

Figure 3.2 – Graphic Resource Settings dialog box

In the dialog box that will open up (see Figure 3.2), select the following:
- **C30/XC16** compiler, this will ensure that the output will be compatible with the syntax of the PIC24 compilers
- **Internal flash**, this will cause the bitmap arrays to be defined as local constant arrays that the linker will place in on chip flash memory.
- **16 bpp**, this is the color resolution of the Mikromedia display

Proceed to click **OK** and close the *Settings* dialog box.

Figure 3.3 – Fingerprint.bmp

The Graphics Resource Converter understands a limited number of input file formats. Note that when it gets to bitmaps files, it accepts only files with the most basic Windows Bitmap format, with a *.BMP* extension and, even then, a limited subset of all the possible combinations of resolutions, color depths and compression methods.
We will review shortly how to handle a much larger variety of inputs and how to operate all the necessary conversions by using additional open source tools.

You can download the *fingerprint.bmp* and many other resource files from the book web site. You can place the bitmap file directly in the main project window, or better create a sub-folder (***images***) where to keep them separate from the source code.

We can now add the bitmap to the GRC project list. Click on the **Image** icon on the toolbar or select **File/Add Images** from the application menu. Navigate to the project folder and then select from the subfolder *images*: **fingerprint.bmp**.

You will notice that the first line of the table inside the Graphic Resource Converter application window is now filled with details about the newly added resource.
This is a good opportunity for us to check that the file we are submitting is being recognized correctly by GRC!

The following information should appear:
- *Compression: none*. Recent versions of the graphics library have started to accept a basic compression format (Run Length Econding or RLE), but we are not using it for this image
- *Type: bitmap*
- *Size: 2068 bytes*. This is the size of the entire bitmap structure on file

- *Description: 108*147 pixels, 1-bits per pixel,* this is a description based on the information retrieved from the file, not simply a comment.

Next, let's choose **Project>Convert** from the main menu or click on the green checkmark icon in the toolbar.

A new dialog box will pop up (see Figure 3.4) to allow us to choose a destination file name and location for the bitmap resource output.

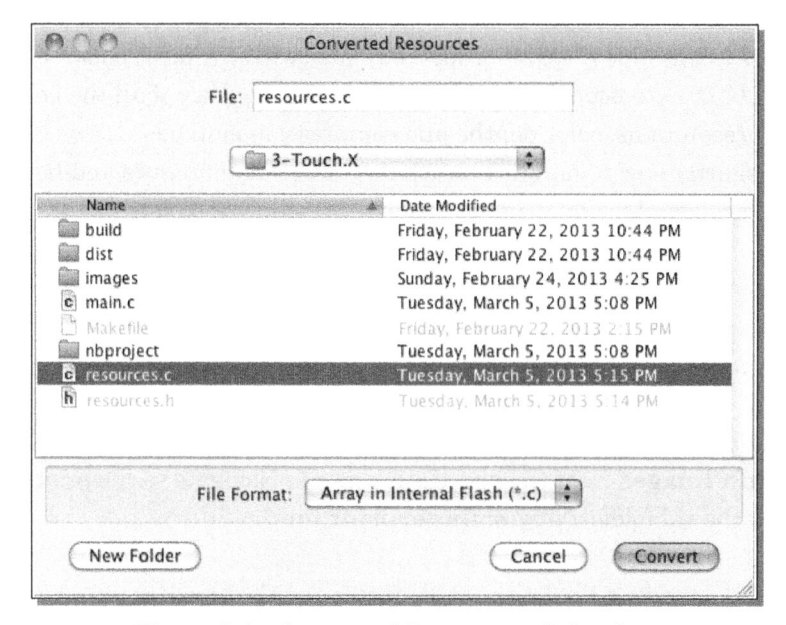

Figure 3.4 – Converted Resources dialog box

Here we will select the project directory and we will type the desired output file name: ***resource.c.***

But wait! Before rushing to click on the *Convert* button, it is very important that you ensure that the *File Format* selection box is set on **Array in Internal flash (*.c)**

BUG ALERT

You might expect the file format to be assigned by default given our current tool configuration. Unfortunately, in my experience with versions up to v3.28, this

> was not the case. Leaving the field empty (the actual default) made for the tool to produce some pretty unpredictable behavior.

The conversion in itself, once you do press the **Convert** button, is a very quick process. Even for a large number of resource files, you should get a rather immediate confirmation of the successful completion – *fingerprint.bmp* has been converted and packed into the file: *resource.c*

A resource file is just a C source file containing the definition of a large constant array of integers. We could have typed it all by hand if we wanted to although, admittedly, it would have been quite a boring exercise even for such a small image!

In the project window, right click on the *Source Files* logical folder and then select **Add Existing Item...** Select the **resources.c** file we just created and confirm.
Now the resource (bitmap) is ready to be linked in with our project application code.

Bitmap Files Under the Hood

At this point, I always like to take a look under the hood. After all, this is still the best way to understand how things work. As for any other source file, we can double click on its name in the Project window and inspect its contents in the Editor window.

Past the usual lengthy file banner and the legal disclaimer, you will find comments (automatically generated by the GRC) separating neatly a few sections containing parameters that will be used later by the library to handle the resources included and eventually (down below) the resources themselves.

In our first example there is only one resource, but a more typical example will see multiple bitmaps (and fonts) all packaged in the same output resource file.

No matter how complex, you can simply perform a search (CTRL+F or ⌘+F) for a label corresponding to the name of the resource you want to observe (*fingerprint*) and place the cursor directly on the definition you want to observe.

```
const IMAGE_FLASH fingerprint =
{
    (FLASH | IMAGE_MBITMAP | COMP_NONE),
    (FLASH_BYTE *)__fingerprint
};
```

Listing 3.1 – Bitmap in flash structure

Listing 3.1 shows in fact how the Graphic Resource Converter has *defined* our bitmap. The IMAGE_FLASH type is defined in *primitives.h* as a structure that contains two fields:

- a set of flags, here it is initialized to say it resides in flash memory, it is a bitmap (as opposed to a font) and it is not compressed
- a pointer to the array proper (`__fingerprint`) which is then initialized, byte by byte, in the following lines

Since the `fingerprint` structure is defined as a *const*, it will be placed in the flash memory of the microcontroller by the MPLAB XC16 linker..

So let's scroll a few lines below to find the definition of the bitmap array `__fingerprint` Notice that what follows is actually inline *assembly* language notation.

```
extern const char __fingerprint[] __attribute__((space(prog), aligned(2)));

asm(".section .text, code");
asm(".global __fingerprint");
asm(".align 2");
asm("__fingerprint:");
```

Listing 3.2 - Bitmap resource array

I am not expecting every reader to be familiar with the notation used in Listing 3.2, but it will suffice to know that the first four lines are basically containing instructions to align the data on word boundaries and to create a label that is accessible to the C compiler and ultimately your application.

```
/******************************************
 * Bitmap header
 ******************************************/
asm(".byte 0x00");              // Compresssion
asm(".byte 0x01");              // Color Depth
asm(".byte 0x93, 0x00");        // Height
asm(".byte 0x6C, 0x00");        // Width
```

Listing 3.3 - Bitmap header

The next section, titled Bitmap header, is where the bitmap data proper starts.
Here, the dimensions of the array (width and height) are provided as 16-bit integers and the color depth is provided. Any color depth that is smaller than the color depth of the display (16-bit in our case) indicates that an indexing system (palette) will be used.

```
/******************************************
 * Color Palette for the image
 ******************************************/
asm(".byte 0x00, 0x00");
asm(".byte 0xff, 0xff");
```

Listing 3.4 - Bitmap color palette (2 colors)

Surely enough the next section is providing the translation via a *color index table*, aka the *palette*. Sequentially each row in this table (corresponding to each line in the source file) will provide the definition of one color (in our case, a black and white picture has only two colors: 0 and 1 corresponding to the 0x0000 and 0xffff words).

```
/*********************************
 * Bitmap Image Body
 *********************************/
asm(".byte 0xFF, 0xFF, 0xFF, 0xFF, 0xFF, 0xDF, 0xFF, 0xFF, 0xFF, 0xFF, 0xFF, 0xFF,
0xFF, 0x0F, 0xFF, 0xFF");
asm(".byte 0xFF, 0xFF, 0x3F, 0x00, 0x00, 0x00, 0xFE, 0xFF, 0xFF, 0xFF, 0xFF, 0x0F,
0xFF, 0xFF, 0xFF, 0xFF");
```

Listing 3.5 - Bitmap image body excerpt

Finally, we'll find the bitmap body itself. Here each byte will represent a group of 8 sequential pixels.
Each bit value is used as an index in the palette table to retrieve the color information that will be transferred on the screen.

Fonts

In the previous chapter we have already made use of another *graphical resource* (*GOLFontDefault.c*) and you might now be wondering if there is any similarity between bitmaps and fonts.
In fact there is a lot of similarity in how the two types of resources are handled and how they are generated (converted). You could think of a font as a collection of small bitmaps, called *glyphs*, one for each character. In *monospaced* font you can expect to find that every character glyph has the same width, while in *proportionally* spaced fonts, each character glyph will have a different width.
Once more, there is nothing better than inspecting an actual font resource file.
Just add the *GOLFontDefault.c* file (found in the *Microchip/Graphics* folder) to the project and open it in the editor window. Then scroll down gradually past the file banner and the legal disclaimer.

Since the *GOLFontDefault.c* file is written generically both for the PIC24 and for the PIC32, you will see that the file is actually split in two halves. Each half is conditionally included when the corresponding compiler model is declared in the project. The PIC24 happens to use the second half. You can scroll down approximately to the middle point or you can simply perform a search for __XC16__, the distinctive symbol used to notify that the PIC24 compiler is in use.

```
const FONT_FLASH GOLFontDefault =
{
    (FLASH | COMP_NONE),
    (GFX_FONT_SPACE char *)__GOLFontDefault
};
```

Listing 3.6 - Font resource in flash struct

Identically to the case of a bitmap resource, we encounter first a structure initialization. The `FONT_FLASH` type is defined in *primitives.h* and contains the exact same two members:

- a set of flags, indicating where the actual data resides (flash) and if there is any compression applied (none)

- and a pointer to the actual data, an array of integers

```
asm(".section .const, psv, page");
#endif
asm(".global ___GOLFontDefault");
asm(".align 2");
asm("___GOLFontDefault:");
```

Listing 3.7 - Font resource array content

The array is initialized in its turn a few lines below, see Listing 3.7, just as in the bitmap resource, using a particular assembly syntax that guarantees the correct alignment of the data in flash memory.

```
/*******************************************
 * Font header
 *******************************************/
asm(".byte 0x00");           // Font ID
asm(".byte 0x00");           // Font information:
                             // 1 bit per pixel, Characters zero degrees rotation
asm(".byte 0x1C, 0x00");     // First Character
asm(".byte 0x7E, 0x00");     // Last Character
asm(".byte 0x1B, 0x00");     // Height
```

Listing 3.8 - Font resource header

A font header follows, which provides a few more bits of information such as:

- *type* of font and additional flags (orientation, number of bit per pixel...)

- *subset* of the character set covered in this resource

- *height* of the font (common to all glyphs)

```
/*********************************
 * Font Glyph Table
 *********************************/
asm(".byte 0x10");                  // width of the glyph
asm(".byte 0x94, 0x01, 0x00");      // Character - 28, offset (0x00000194)
asm(".byte 0x10");                  // width of the glyph
```

Listing 3.9 - Font Glyph index table excerpt

Next is the font glyph (index) table. Since each glyph is potentially of a different size (in proportional fonts), this table specifies for each one the size and relative offset in the *characters* table. Thanks to the glyph index table it is possible to pack each glyph (bitmap) in memory more efficiently, without any padding, reducing considerably the overall memory usage.

As you keep scrolling down into the character table, you will now see one by one, each character bitmap. Larger characters will use two or more bytes per row, while narrow characters might use only one byte per row.

```
/*********************************
 * Font Characters
 *********************************/
...
/*********************************
 * Character - 49
 *********************************/
asm(".byte 0x00");      //
asm(".byte 0x1C");      //    ***
asm(".byte 0x1E");      //    ****
asm(".byte 0x18");      //    **
asm(".byte 0x18");      //    **
asm(".byte 0x18");      //    **
asm(".byte 0x18");      //    **
asm(".byte 0x18");      //    **
asm(".byte 0x3E");      //    *****
asm(".byte 0x00");      //
asm(".byte 0x00");      //
asm(".byte 0x00");      //
```

Listing 3.10 - Glyph for character 0x49

Courtesy of the GRC tool, used to generate the *GOLFontDefault.c* resource, each row of data in the *Characters* table has also a convenient depiction in the form of a comment using the * symbol to represent each pixel, a la *ASCII art*!

Having satisfied our curiosity and having learned quite a bit in the process, it is time to put bitmap and font resources to use in a simple application!

First Demo Project: Splash Screen

As a first project, we will set off to design a simple splash screen. We'll put a title (banner) at the top of the screen and then display a bitmap (fingerprint) well centered below it.

```c
/*
 * File:    main.c
 *
 * Project: 3. Splash screen demo
 *
 */
#include "PICconfig.h"
#include "Graphics/Graphics.h"

#include "resources.h"

int main( void )
{
    int width;
    char s[] = "Fonts and Bitmaps";

    // 1. init
    InitGraph();                    // init graphics library
    SetColor( WHITE );              // set background color
    ClearDevice();                  // clear display contents
    DisplayBacklightOn();           // turn on the backlight

    // 2. display centered banner
    SetColor( BRIGHTRED );          // set color
    SetFont( (void*) &GOLFontDefault);
    width = GetTextWidth( s, (void*) &GOLFontDefault);
    OutTextXY( (GetMaxX()-width)/2 , 0, s);

    // 3. display centered bitmap
    width = GetImageWidth( (void*) &fingerprint);
    PutImage( (GetMaxX() - width)/2, 60, (void*) &fingerprint, IMAGE_NORMAL);

    // main loop
    while( 1)
    {

    } // main loop
}
```

Listing 3.11 - main.c Splashscreen demo

The code in Listing 3.11 takes care of centering horizontally both the text banner and the bitmap. Key to properly align text and images on the display is knowing how to obtain the size of each graphical element. If you worked through the exercises in the previous chapter (or skipped to the solutions) you will have learned that the width in pixels of a string of text (as it would be rendered on the screen for a given proportional font) can be had by using the `GetTextWidth()` function. An equally handy function is available to obtain the width (and height) of any bitmap: `GetImageWidth()`. The use of this function is highly recommended even if you happen to know the dimensions of your bitmaps ahead of time.

The time required to fetch the value from the FLASH_IMAGE structure is negligible (a few cycles, sub-microsecond) when compared to the time actually spent reproducing the image on the screen (tens of thousands of cycles, milliseconds). This gives us the peace of mind of always using the correct values so that future project updates won't *break* the graphics (alignment) should the bitmaps be edited and resized.

Going Large!

As in the first chapter, I could let you go ahead and try and compile the project as is, but I will spare you the drama this time and warn you ahead of time of a memory allocation problem you would have run into. By adding both a font resource and a bitmap resource we have increased our project program memory usage just past the point where the compiler cannot assume anymore that all functions will be accessible as *near* using the default Small Memory Model.

Proceed to open the **Project Properties** dialog box, click on the **x16-gcc** Category, select **Memory Models** among the Options Categories and finally select the **Large** code model!

Once the project builds successfully, load it onto the Mikromedia board and you should see a monochromatic splash screen image appear as in Figure 3.5

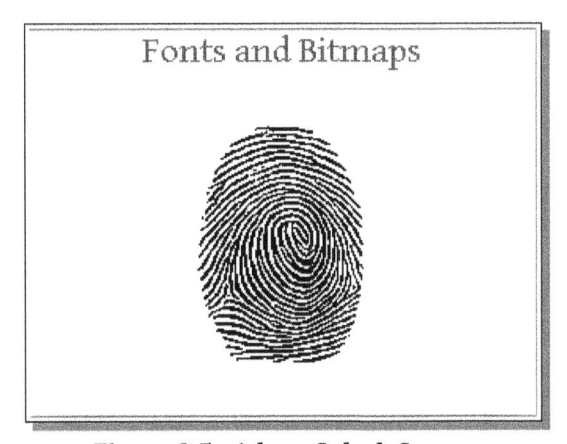

Figure 3.5 - 1-bpp Splash Screen

A Colorful Splash Screen

Changing from 1-bpp black and white bitmaps to 4, 8 or 16-bpp colorful bitmaps does not change the complexity of the project.

Using the GRC tool, we can pack multiple bitmaps, mixing and matching sizes and color depths as needed, simply producing a proportionally larger *resource.c* output file.

From the book web site, download the *flower.bmp* (a 16bpp example) and add it to the Graphic Resource Converter project (*table of resources*) to generate a new output *resource.c* file. Instead of centering a single bitmap on screen, this time we will show two images side by side. Since the same `PutImage()` function is perfectly capable of handling all color depth resources, we can simply replace the last few lines of the main.c file used in the previous example project with:

```
// 3. display two bitmaps with padding
w1 = GetImageWidth( (void*)&fingerprint);
w2 = GetImageWidth( (void*)&flower);
pad = (GetMaxX()- w1 - w2) / 3;
PutImage( pad, 60, (void*) &fingerprint, IMAGE_NORMAL);
PutImage( pad + w1 + pad, 60, (void*) &flower, IMAGE_NORMAL);
```

This time, a little extra code was added to ensure that an equal amount of padding is added in between the bitmaps and on the sides. Take care of adding the declarations of the new variables `w1`, `w2` and `pad` as `unsigned` or the MLA own `SHORT` type.

Rebuild and launch the project and observe how different is the result of the two function calls side by side.

Figure 3.6 - A 16-bpp Splash Screen

Terminal Emulation

As nice as the graphics capabilities of the Mikromedia display are, you will soon find yourself wishing you had a simple alphanumeric display to write to.
Many demo projects in the MLA libraries, for example, are based on the assumption that there is available an alphanumeric LCD display (as the one featured on the Explorer 16 board). Wouldn't it be nice to be able to run all those demos (with minimal modifications) on the Mikromedia board too?
But there is another reason. When debugging even the most complex projects, the simplicity and immediateness of the good old *printf()* to log intermediate values of a calculation onto a console is hard to beat.

So in the rest of this chapter, we will create a minuscule terminal emulation module that will mimic in part an alphanumeric LCD display and in part the text page of a traditional computer screen.
On the Mikromedia board display we will simulate a text page by using a monospaced font or by forcing a proportional one to use a fixed amount of space per character.
By choosing a fixed font size of 12*20 pixel for example, we will be able to get a (landscape) text page of 12 lines by 26 characters.

For maximum compatibility we will conform to the API of the LCD display driver used by many Explorer16 demo projects. Here the few functions that we will need to implement:

```
void LCDInit(void);
void LCDHome(void);
void LCDL2Home(void);
void LCDClear(void);
void LCDPut( char);
void LCDPutChar( char);
void LCDPutString( char *);
```

About Scrolling Text

Things will get slightly more challenging when we attempt to add emulation of a proper terminal *scrolling* functionality.
Scrolling the Mikromedia display pixel by pixel can be a lengthy and inefficient process. Writing to each screen pixel via the PMP port (in 8-bit mode) requires two consecutive write commands, but reading the same pixel color information back involves many more cycles.
Since we are only interested in scrolling text contents, it is much more efficient to keep a

small table inside the PIC24 RAM to store each character that has been rendered on the screen. Note that we are only storing the characters, not the pixel information. Upon a scroll request, we can perform a quick scroll on this table, by moving only 12*26 bytes of character data and then rendering again the entire screen.

This technique can significantly increase the efficiency of the scroll operation, but comes at the cost of RAM, a relatively precious resource. Fortunately, in most applications we will find that the 312 bytes of memory necessary to hold the text *page* are a relatively small price to pay when compared to the total amount of RAM available to the PIC24 and PIC32 microcontrollers used on the Mikromedia boards (16K and 128K respectively).

Further, we can make the scrolling functionality optional by implementing a simple configuration file similar to the one used by the Graphics library (*LCDConfig.h*) to enable or disable this and other features as necessary:

- `LCD_SCROLL`, if defined, will allocate space in RAM for the purpose, enabling automatic scrolling once the cursor has reached the bottom of the screen

Similarly we can control *wrapping* of the cursor once it reaches the end of a line.
This would be considered normal behavior for a computer terminal, but it would otherwise break the emulation of an alphanumeric LCD display.

- `LCD_WRAP`, if defined, allows the cursor to wrap around onto a new line once the cursor has reached the right edge of the present one.

Finally since it can be annoying to constantly take care of switching colors used by the graphics primitives and we do not want that to interfere with the terminal emulation, we can define two properties (static variables) to record the currently selected color for text background and foreground. These can be initialized by adding two more symbols to the *LCDConfig.h* file:

- `LCD_FORE`, defines the default text foreground color
- `LCD_BACK`, defines the default text background color

With these premises in mind, you will appreciate the simplicity of the small library module, that we will call: *Terminal.c* and its corresponding *Terminal.h* header file. These files can be conveniently added to the *uMB* folder (created in the previous chapter) inside the working directory.

In the following few pages, we will review the essential portions of this module, starting from a small set of *setters* and *getters* functions.

```
/*********************************************************************
 *
 *      Terminal Emulation on Mikromedia LCD display
 *
 *      API compatible with Explorer16 2x16 LCD interface
 *********************************************************************
 * File Name:           LCDterminal.c
 *********************************************************************/
#include <string.h>
#include "LCDTerminal.h"

extern const FONT_FLASH TerminalFont;

static int _cx, _cy;      // cursor position
static int _back = LCD_BACK , _fore = LCD_FORE;

#ifdef LCD_SCROLL
    // text page char matrix
    static char page[ _MAX_Y][ _MAX_X+1];
#endif

int LCDGetX()
{
    return _cx;
}

int LCDGetY()
{
    return _cy;
}

void LCDSetColor( int x)
{
    _fore = x;
}

void LCDSetBackground( int x)
{
    _back = x;
}

int LCDGetColor()
{
    return _fore;
}

int LCDGetBackground()
{
    return _back;
}
```

Listing 3.12 - LCDterminal.c

As anticipated, a bi-dimensional array called *page* was defined to contain the ASCII characters placed on the "text page" only if the scroll switch is defined. Since it is defined as *static*, it won't be accessible to functions outside this module source file.

```c
void LCDHome(void)
{   // set cursor to home position
    _cx = 0;
    _cy = 0;
}

void LCDSetXY( int x, int y)
{
    _cx = x; _cy = y;
}

#define LCDL1Home    LCDHome

void LCDL2Home(void)
{
    _cx = 0;
    _cy = 1;
}

void LCDLineHome(void)
{
    _cx = 0;
}
```

Listing 3.13 - LCDterminal.c (continued)

These are functions that take care of the basic cursor positioning and are borrowed from the alphanumeric LCD library API.

```c
void LCDClearToEOL( void)
{
    int t = _cx;
    int i;

    // fill with spaces until the end of the line
    for( i=_cx; i< _MAX_X; i++)
        LCDPut( ' ');

    // return to position
    _cx = t;
}

void LCDClear(void)
{
#ifdef LCD_SCROLL
    int i,j;
    for (i=0; i< _MAX_X; i++)
        for( j=0; j<_MAX_Y; j++)
            page[j][i] = ' ';
#endif

    SetColor( _back);
    ClearDevice();
    SetColor( _fore);
    LCDHome();
}

void LCDInit(void)
{
    InitGraph();    // initialize graphics library
    LCDClear();
    SetFont( (void *)&TerminalFont);        // set font
}
```

Listing 3.14 - LCDterminal.c (continued)

`LCDInit()`, `LCDClear()` and `LCDClearToEOL()` are taking care of the graphics library initialization, color selection and eventually the initialization of the *page* in case the scroll switch is defined.

Note that during the initialization a specific font resource is referenced (`TerminalFont`). This will require that we include it in the resource file (using GRC) or as a separate entity.

```c
void LCDShiftCursorLeft(void)
{
    if (_cx>0)  _cx--;
}


void LCDShiftCursorUp(void)
{
    if ( _cy>0) _cy--;
}

void LCDShiftCursorDown(void)
{
    // advance to next line
    _cy++;

#ifdef  LCD_SCROLL
    if ( _cy >= _MAX_Y)
    {   // scroll entire screen up
        int i,j;
        for( j=0; j<_MAX_Y-1; j++)
        {   // for each line
            // clear the line background (independent from LCD_OVERLAY)
            SetColor( _back);
            Bar( 0, FONT_H*j, GetMaxX()-1, FONT_H*(j+1));
            SetColor( _fore);
            // copy from next line print the new line content
            for( i=0; i<_MAX_X; i++)
            {
                page[j][i] = page[j+1][i];
                MoveTo( FONT_W*i, FONT_H*j); OutChar( page[j][i]);
            }
        }

        // limit to last line
        _cy = _MAX_Y-1;

        // clean up the last line
        SetColor( _back);
        Bar( 0, FONT_H*_cy, GetMaxX()-1, FONT_H*(_cy+1));
        for( j=0; j<_MAX_X; j++) page[_cy][j]=' ';
        SetColor( _fore);

    }
#else   // no scrolling option roll
    if ( _cy >= _MAX_Y)
    {
        _cy = 0;
    }

#endif
}
```
Listing 3.15 - LCDterminal.c (continued)

```c
void LCDShiftCursorRight(void)
{
    _cx++;
#ifdef LCD_WRAP
    if ( _cx >= _MAX_X)
    {   // wrap to a new line
        _cx = 0;
        LCDShiftCursorDown();
    }
#endif
}
```
Listing 3.16 - LCDterminal.c (continued)

The `LCDShift`**xxx**`()` function group is taking care of the cursor as it moves in the four directions according to the following simple rules:

- Going up, there is a hard ceiling. There is no provision for a back-scroll as there is no logging of the text page content beyond the actual page rendered.
- Going left means similarly hitting a wall when the cursor reaches the left margin (column 0). This is expected behavior in a console.
- Going down can produce a scroll up of the entire page once the bottom is reached (if the scroll symbol is defined)
- Going right can produce a wrap onto the next line (if the wrap symbol is defined), which in turn can mean a scroll up of the entire page if the current line is at the bottom of the screen (and the scroll symbol is defined)

```
void LCDPut(char A)
{
#ifdef LCD_SCROLL
        page[ _cy][ _cx] = A;
#endif

#ifndef LCD_OVERLAY
    // clear the background
    SetColor( _back);
    Bar( _cx*FONT_W, _cy*FONT_H, (_cx+1)*FONT_W, (_cy+1)*FONT_H);
    SetColor( _fore);
#endif

    if(( _cx<_MAX_X) && (_cy<_MAX_Y))    // clip
    { // print the new character
        MoveTo( FONT_W * _cx, FONT_H*_cy);
        OutChar( A);
        LCDShiftCursorRight();
    }
} // LCDPut
```

Listing 3.17 - LCDterminal.c (continued)

LCDPut() is the function that ultimately places each character to the screen and does so in a monospaced grid regardless of the font spacing properties. The grid is based on a pre-defined fixed character height and width (`FONT_W` and `FONT_H`).

It considers the scroll switch and takes care of recording each ASCII character in the *page* array, but it introduces also a new option and a new switch, `LCD_OVERLAY`, to activate it. In fact the graphics primitives `OutText()` and `OutChar()`, when rendering a character, simply over-impose the bitmap (glyph) on top of any preexisting image (or character) present in a given position. In many cases, this is not what you want. An opaque background is sometimes helpful in making text more readable and definitely desirable in a more traditional console emulation.

So by default, the terminal module does clear the rectangular area (using the background color) first and then renders a character (using the foreground color) on top of it. When the *overlay* switch is defined instead, the background is not cleared and text is simply added on top of any preexisting color information.

```
void LCDPutChar(char A)
{
    int tab, i;

    switch( A)
    {
      case '\b':    // backstep
        if ( _cx>0)
        {
            _cx--;
            LCDPut(' ');
            _cx--;
        }
        break;

      case '\t':    // move to next tab position
        tab = (_cx/8 + 1) * 8;
        // add spaces until the next tab stop
        for( i=_cx; i<tab-1; i++)
            LCDPut(' ');
        break;

      case '\n':    // New Line
#ifndef LCD_OVERLAY           // clear rest of the line
        SetColor( _back);
        Bar( FONT_W*_cx, FONT_H*_cy, GetMaxX()-1, FONT_H*(_cy+1));
        SetColor( _fore);
#endif
        LCDShiftCursorDown();
        // break;    // continue into Home
      case '\r':    // Home
        _cx = 0;
        break;

      default:      // print-able char
        LCDPut( A);
        break;
    } // switch

} // LCDPutChar
```

Listing 3.18 - LCDterminal (continued)

`LCDPutChar()` is built on top of `LCDPut()` and takes care of interpreting special characters such as the *newline, carriage return* and *tab* as you would expect in a basic terminal emulation. It also performs a special handling of the *backspace* character, which provides the ability to perform basic editing of a command line on the console.

```
void LCDPutString( char* s)
{
    char c;
    while( (c = *s++))
        LCDPutChar( c);
}
```

Listing 3.19 - LCDterminal.c (continued)

`LCDPutString()` is trivially a translation of the C standard I/O library function `puts()` and, since it is built on top of `LCDPutChar()`, inherits the special characters handling too.

```
void LCDCenterString( int p, char *s)
{   // p  integer offset (lines) above or below center
    // s  string
    int x, y;

    // get center position
    y = _MAX_Y/2 -1 + p;
    x = (_MAX_X - strlen( s))/2;

    // set and print
    LCDSetXY( x, y);
    LCDPutString( s);
} // LCDCenterString
```

Listing 3.20- LCDterminal.c (continued)

`LCDCenterString()` is built on top of `LCDPutString()`, but adds the ability to center horizontally the string of text on the screen. There is though an additional parameter passed to this function (p) which makes it very efficient at centering multiple strings on the screen both horizontally and vertically.

You can think of this parameter as a vertical positioning offset. If you pass the value zero, the string will be centered both horizontally and vertically.

```
LCDCenterString( 0, "Mikromedia");
```

If you pass a positive integer (+1, +2...+n) the string will be centered horizontally but will be placed a given number of lines below the center of the screen. Negative values will displace the string a given number of lines above the center of the string.
In other words, in homage to my laziness, this function can be used to quickly create a splash screen of sorts with minimal coding and no manual calculations.

```
LCDCenterString( -1, "Title of the Demo");
LCDCenterString(  0, "+");
LCDCenterString( +1, "Touch the screen to Start");
```

Sharing the Terminal Module

To use the terminal emulation module we will need to place the *LCDterminal.c* and *LCDTerminal.h* files in the *uMB* folder inside the working directory. In the same folder we will store the *LCDTerminalFont.c* file that contains the hand optimized monospaced font resource (`TerminalFont`), after downloading it from the book web site.

All projects that will want to use the LCD Terminal library and its resources will be able to do so provided the *uMB* folder is included in their ***C include dirs*** lists (part of the Project Properties) and ensuring that all three files are included in the *Source Files* logical folder of the project.

Summary

As we have seen, font and image (bitmap) resources can be easily included in our projects to expand our visual interface capabilities. We learned to position accurately images on the screen, and how to select a font to render and place text on the screen.
With help from the GRC tool, we learned how to package multiple such resources in a source file for convenient and immediate access from the on chip flash memory of the microcontroller.

Finally, we developed a simple Terminal Emulation module (that we added to our Board Support Package folder, *uMB*) to provide us with convenient text page access to the display of the Mikromedia board. This will come in handy for several future projects helping us test and debug our applications.

Tips & Tricks

Creating bitmap images for use with the Graphics library requires some planning and preparation work. The Mikromedia display color rendering capabilities can differ significantly from those of your personal computer screen for example.

Cropping and resizing images to fit your user interface needs can require additional experimentation and can be a bit of an iterative process.

For this purpose, among the many graphical tools available, I strongly recommend the *GIMP*. This is an open source application available on all major desktop operating systems, and most importantly, capable of producing a bitmap format that is compatible with the Graphic Resource Compiler.

Figure 3.7 – GIMP Spash Scren

NOTE for WINDOWS Users

Do NOT trust the native graphical application (Paint.exe) to produce a valid .bmp file. It will generate a 32-bit bitmap format instead that is incompatible with the GRC tool and the MLA library (besides doubling the memory usage).

Once an image has been edited and is ready to be saved, remember that GIMP, just like Photoshop and most similar applications, uses a proprietary format to store a high-resolution loss-less version of the project.

To produce a standard .bmp or .jpg file output, use the **Export** function instead. In the *Export* dialog box, click on the **Select File Type** option (bottom left) and make sure to select the "**Windows BMP image**" format.

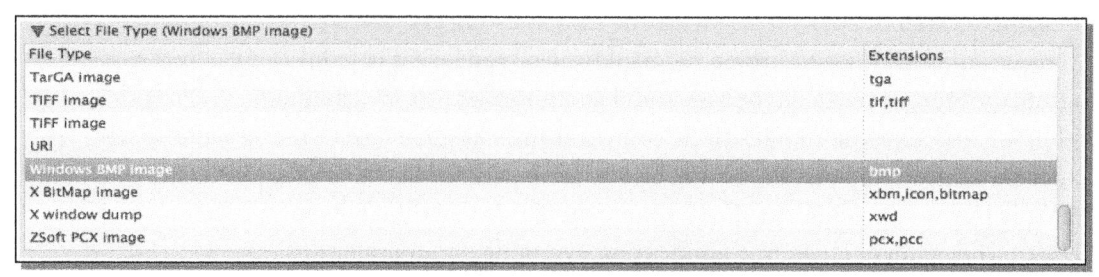

Figure 3.8 - GIMP, Export dialog box

Online Resources

- http://www.flyingpic24.com

    You can find here links to all the resources needed for this and previous books, including code repositories, errata and updates

- http://www.gimp.org
    This is the GNU Image Manipulation Program project home page

- http://www.gimp.org/docs/
    Manuals are available in multiple languages, and the tutorials cover all levels of complexity.

Suggested Reading

- AN1182 – Fonts in the Microchip Graphics Library

    An introduction to font resources optimization

- Wikipedia – Typeface

    To learn more about digital and non fonts and resources

- Google – Fonts

    Hundreds of fonts, free to download and convert

Exercises

1. Use GIMP to "format" an image as "indexed" (to use color bitmaps) and export multiple versions of the same image with different bit depths: 2-bit (black and white), 4-bit (16 shades), 8-bit (256 colors). Compare the images side by side using the second demonstration project developed in this chapter.

2. Redefine the system `write()` function so that `printf()` output is automatically redirected to the LCD terminal emulator.

    You can add this function to the LCD Terminal module so that in all future projects you will be able to use directly `printf()` in place of `LCDPutString()`!

Solution to Exercise 2

```c
/*
 * Project: 3.Solution -printf
 *
 * File:    main.c
 *
 */
#include "PICconfig.h"
#include "LCDTerminal.h"

#include <stdio.h>          // sprintf
#include <stdarg.h>         // var args

void Log( const char *fmt, ...)
{
    char s[128];
    va_list argp;

    va_start( argp, fmt);
    vsprintf( s, fmt, argp);
    LCDPutString( s);
    LCDPutChar( '\n');
    va_end( argp);
}

int write( int handle, char *p, unsigned len )
{
    unsigned i = len;

    while ( i-- > 0)
        LCDPutChar( *p++);

    return len;
}

int main( void )
{
    int i;

    // init the graphics
    LCDInit();
    DisplayBacklightOn();

    // main loop
    while( 1 )
    {
        //Log( "Hello World #%04x", i++);
        printf( "Hello World #%04x", i++);
        //DelayMs(20);

    } // main loop
} // main
```

Listing 3.21 - Solution to Exercise 3.2

Chapter 4

Touch Input

In this chapter, we will move on to the other side of the user interface equation. In fact up to this point, we have been mostly involved in developing ways to produce a visible *output*. We will now explore ways to enable our Mikromedia board to receive user *input by* introducing *touch sensing* and by taking advantage of the resistive touch screen available on the Mikromedia boards.

Let's take a second look at the board schematics and, in particular, at the section detailing the display interface. We can now focus on the bottom four connections of the TFT display module: pins 44, 45, 46, 47 are in fact providing us with access to the two layers of a resistive touch screen.

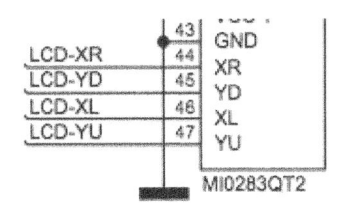

Figure 4.1 - Mikromedia schematic - Touch Screen connections detail

Touch Technologies

It is not my intention to provide a detailed summary of all touch sensing technologies here, but a short overview is in order.

The two main classes of touch technologies available to embedded control applications are: *resistive* and *capacitive*. Each one comes with its own set of pros and cons.

Perhaps the biggest distinction is that resistive touch sensing solutions are capable of detecting at most one point of contact (*single-touch*) while capacitive solutions can detect multiple points of contact (*multitouch*). Resistive touch screens can be operated with styluses, fingers and even when wearing thick gloves. Capacitive touch screens, on the contrary, can be much more particular and require a naked finger or other "capacitive" pointer.

Resistive screens require more pressure than capacitive ones, but while this might appear to be a negative aspect, it is balanced by the resulting increased resolution achievable.

Resistive touch *sensing* can be performed very easily by *any* microcontroller by using a minimum of four pins of which only two are analog inputs. An Analog to Digital converter peripheral of medium resolution (10-bit) will do the job.

By contrast capacitive touch sensing requires a more complex, dedicated, analog interface (more pins, more inputs, higher voltages, higher sensitivity to noise) and a significantly higher workload for the microcontroller, increasing with the number of touch points recognized and possibly additional requirements for gesture detection.

Not only does this mean that a custom controller is often required to take care of capacitive touch screens, but the cost of the solution is intrinsically higher as the transparent conductive layer materials used (ITO) are significantly more expensive than the materials used in resistive screens.

Hopefully this gives you an appreciation for the compromises that have been made by the Mikromedia board designers who favored a simpler but effective input mechanism for embedded control applications over the more flashy but perhaps unnecessary multi-touch capacitive alternative common in so many consumer products (tablets and cell phones).

Elements of a Resistive Touch Screen

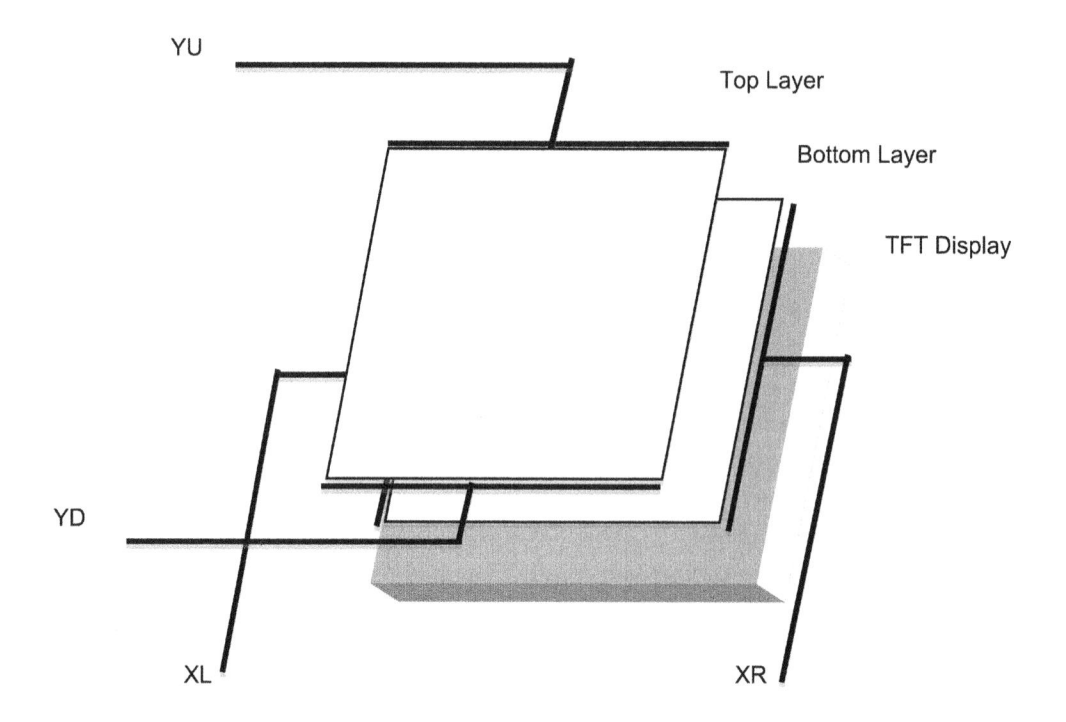

Figure 4.2 - Simplified structure of a Resistive Touch Screen

The simplified physical structure of a (4-wire) resistive touch screen is illustrated in Figure 4.2. Two resistive (transparent) layers are separated by a small air gap. When there is no pressure applied, there is actually no electrical contact. The electrodes (conductive bars) are connected at the sides of each layer. When a voltage differential is applied, the resistive nature of the material produces a linear gradient along one axis. By having the electrodes connected on the two layers in an orthogonal fashion, a gradient can be produced on both the X and the Y-axis alternatively polarizing the electrodes on one layer and using the other layer as a pick up point.

In fact, by changing the configuration of the I/O port pins to which the two layers electrodes are connected, we can obtain a number of different measurements. For example we can get:

1. *Touch detection*, by setting YD as a digital output (low), and XR as an analog input

2. *X coordinate measurement*, by setting XR and XL as digital output pins (high and low respectively), and YU used as an analog input

3. *Touch confirmation*, by setting XL as a digital output (low), and YU used as an analog input

4. *Y coordinate measurement*, by setting YU and YL as digital output pins (high and low respectively), and XR used as an analog input

A simple state machine can be made to rotate through the four states. Actually the resulting state machine is going to look more like a 9 state sequence. The extra states provide for a stabilization delay, in order to avoid measurement errors due to parasitic (capacitive) effects of the two plates and parasitic (inductive) effects of the traces and electrodes.

State	Description	Next	If fails
Set X	Configure I/Os, start conversion	Check X	
Check X	Check if pressure detected	Sample X	Set X
Sample X	Reconfigure I/Os, start new conversion	Get X	
Get X	Record X coordinate (temp X)	Set Y	
Set Y	Configure I/Os, start new conversion	Check Y	
Check Y	Check if pressure still applied	Sample Y	Set X
Sample Y	Reconfigure I/Os, start new conversion	Get Y	
Get Y	Record Y coordinate (temp Y)	Publish XY	
Publish XY	Check if pressure is still applied, update X, Y values with temps	Set X	Set X

Table 4.1 – Touch Sensing State Machine - transition table

Note that extra care must be taken to verify that the (touch) pressure is not removed while the X and Y values are being obtained. That's why we are so carefully checking that pressure is applied shortly before and after each axis measurement and before publishing the new X,Y pair values.

Board Support Package – TouchScreen.c

Resistive touchscreen support has been part of the Graphics library since the very early revisions, but has been mostly relegated to hardware specific modules bundled inside demonstration projects. In recent revisions of the library, these and many similar hardware support modules (accelerometer, beeper, EEPROM, Serial Flash...) have been conveniently grouped inside the MLA *Board Support Package* folder.

Support for touchscreens has been expanded to include interfaces to standalone controller devices (such as the AR1020 series) and the door has been left open for future releases to include *projective capacitive* solutions too.

The touchscreen module is therefore now split into two separate C source files and their respective header files:

- *TouchScreen.c*, generalizes access to the touchscreen interface providing generic initialization, calibration, non volatile storage and graphic object layer message generation (which will be covered in a later chapter when discussing the Graphics Object Library)
- *TouchScreenResistive.c*, is the module doing the actual heavy lifting. It includes an initialization function `TouchInit()`, and a state machine (similar to the one described in Table 4.1) available as the `TouchDetectPosition()` function.

NOTE

> Looking for touch support in the Microchip Libraries for Applications, we must not confuse the *mTouch™* library (*Microchip/mTouchCap* folder) with the actual touchscreen support. *mTouch* is a trademark of Microchip Technology Inc. and refers generically to the *capacitive* touch solutions used to replace mechanical buttons and sliders in consumer and embedded control applications.

Configuring the Touchscreen

There are a significant number of symbols that must be properly defined both in the Hardware Profile and the Graphics Configuration files. Let's review them gradually so that we don't get lost in the labyrinth.

Starting with a local fresh copy of the *GraphicsConfig.h* file, let's add (or uncomment if already present) the line containing:

```
#define USE_TOUCHSCREEN
```

This is found in the *input selection* section of the standard template.
A few lines below, in the *font selection* section, let's edit the line containing:

```
#define FONTDEFAULT        TerminalFont
```

This will ensure that the font we have selected for our LCD terminal emulation (monospaced) will also be used by the touchscreen calibration routines avoiding the need to have a second font added to the project only for this purpose.

The edits to the *HardwareProfile.h* file are more consistent as we need to add an entire new section to it. Let's proceed in small steps, appending the following lines of code:

```
/*******************************************************************
 * IOs for the Touch Screen
 *******************************************************************/

#define USE_TOUCHSCREEN_RESISTIVE

#define TOUCH_ADC_INPUT_SEL    AD1CHS

// ADC Sample Start
#define TOUCH_ADC_START        AD1CON1bits.SAMP

// ADC Status
#define TOUCH_ADC_DONE         AD1CON1bits.DONE
```

Listing 4.1 - Hardware Profile Edits

This very first section defines the type of touchscreen available physically on the Mikromedia board (resistive) and then proceeds to define three macros that help abstract the microcontroller ADC peripheral: channel selection register, sampling control bit and conversion completion flag.

```
// Analog inputs definitions
#define ADC_XPOS               13
#define ADC_YPOS               12

#define ADPCFG_XPOS            AD1PCFGbits.PCFG13
#define ADPCFG_YPOS            AD1PCFGbits.PCFG12

#define RESISTIVETOUCH_ANALOG  0
#define RESISTIVETOUCH_DIGITAL 1
```

Listing 4.2 - Hardware Profile Edits (continued)

This segment is responsible for the definition of the pins used for the X and Y analog inputs into the ADC input multiplexer.

```
// X port definitions
#define ResistiveTouchScreen_XPlus_Drive_High()       LATBbits.LATB13   = 1
#define ResistiveTouchScreen_XPlus_Drive_Low()        LATBbits.LATB13   = 0
#define ResistiveTouchScreen_XPlus_Config_As_Input()  TRISBbits.TRISB13 = 1
#define ResistiveTouchScreen_XPlus_Config_As_Output() TRISBbits.TRISB13 = 0

#define ResistiveTouchScreen_XMinus_Drive_High()       LATBbits.LATB11   = 1
#define ResistiveTouchScreen_XMinus_Drive_Low()        LATBbits.LATB11   = 0
#define ResistiveTouchScreen_XMinus_Config_As_Input()  TRISBbits.TRISB11 = 1
#define ResistiveTouchScreen_XMinus_Config_As_Output() TRISBbits.TRISB11 = 0

// Y port definitions
#define ResistiveTouchScreen_YPlus_Drive_High()       LATBbits.LATB12   = 1
#define ResistiveTouchScreen_YPlus_Drive_Low()        LATBbits.LATB12   = 0
#define ResistiveTouchScreen_YPlus_Config_As_Input()  TRISBbits.TRISB12 = 1
#define ResistiveTouchScreen_YPlus_Config_As_Output() TRISBbits.TRISB12 = 0

#define ResistiveTouchScreen_YMinus_Drive_High()       LATBbits.LATB10   = 1
#define ResistiveTouchScreen_YMinus_Drive_Low()        LATBbits.LATB10   = 0
#define ResistiveTouchScreen_YMinus_Config_As_Input()  TRISBbits.TRISB10 = 1
#define ResistiveTouchScreen_YMinus_Config_As_Output() TRISBbits.TRISB10 = 0
```

Listing 4.3 - Hardware Profile Edits (continued)

The section in Listing 4.3 is abstracting the I/O pins configuration (input/output) and the X and Y electrodes configurations required for proper pressure detection and coordinate reading.

```
// serial Flash calibration addresses
#define ADDRESS_RESISTIVE_TOUCH_VERSION  (unsigned long)0xFFFFFFFE
#define ADDRESS_RESISTIVE_TOUCH_ULX      (unsigned long)0xFFFFFFFC
#define ADDRESS_RESISTIVE_TOUCH_ULY      (unsigned long)0xFFFFFFFA
#define ADDRESS_RESISTIVE_TOUCH_URX      (unsigned long)0xFFFFFFF8
#define ADDRESS_RESISTIVE_TOUCH_URY      (unsigned long)0xFFFFFFF6

#define ADDRESS_RESISTIVE_TOUCH_LLX      (unsigned long)0xFFFFFFF4
#define ADDRESS_RESISTIVE_TOUCH_LLY      (unsigned long)0xFFFFFFF2
#define ADDRESS_RESISTIVE_TOUCH_LRX      (unsigned long)0xFFFFFFF0
#define ADDRESS_RESISTIVE_TOUCH_LRY      (unsigned long)0xFFFFFFEE
```

Listing 4.4 - Hardware Profile Edits (continued)

Finally Listing 4.4 defines addresses to be used in conjunction with a non-volatile memory to store/retrieve touchscreen calibration values.

> **NOTE**
>
> We can save the edited Hardware Profile in the shared folder *(uMB)*, as these defaults will be useful in most/all future projects. Graphics Configuration changes, particularly the one affecting the default font selection, should be kept in the project directory so to affect this project alone.

First Touch

Let's create a first demonstration project to put the touch screen interface to test. Thanks to the terminal emulation module we will log directly on the screen the precise coordinates of each touch event detected.

Preparation

Continuing from where we left at the end of Chapter 3, we will create a new project and immediately populate it with the key components of the Graphics library, LCD terminal and touch screen modules. Here is the detailed sequence of steps:

1. Use the *New Project Wizard* to create a new project inside the working directory (*Mikromedia*), let's call it: "**4.1-Touch.X**"
2. Add a new logical folder to the Source Files folder, called **MLA**
3. Add the following items to the folder:
    - *HX8347.c or IL9341*, the display controller driver
    - *Primitive.c* , the primitive layer of the graphics library
    - *TimeDelay.c*, a few basic timing (blocking) functions used in the driver
4. Add a new logical folder to the Source Files folder, called **uMB**
5. Add the following items to the folder:
    - *LCDTerminal.c*, terminal emulation module
    - *LCDTerminalFont.c*, monospaced font resource
    - *TouchScreen.c*, generic touchscreen support module
    - *TouchScreenResistive.c*, resistive touchscreen support module
6. Configure MPLAB® X "*C Include dirs*" to contain:
    - . (dot), the current project directory
    - *../Microchip/Include*, for our source files to reach inside the MLA
    - *../uMB,*

Create a new *main.c* file using the **New File** wizard and the embedded template or your own customized version.

Touch Demo

First we will need to include the usual device configuration, the terminal emulation and the touch screen header files:

```
/*
 * Project: 4.1 Touch
 * File:    main.c
 *
 * Requires: MLA 1306
 */
#include "PICconfig.h"
#include "LCDTerminal.h"
#include "TouchScreen.h"

#include <stdio.h>          // required by sprintf
```

Listing 4.5 - Touch Demo

Note that we are also including the standard C I/O library as we will be using the `sprintf()` function shortly.

Next we'll define an interrupt service routine that, connected to Timer 3 (my personal choice), will ensure the periodic activation of the touch state machine:

```
#define PERIOD_MS    1           // 1ms
#define _ISR  __attribute__((interrupt, shadow, no_auto_psv))

void _ISR _T3Interrupt( void)
{
    _T3IF = 0;
    TouchDetectPosition();
}
```

Listing 4.6 - Touch Demo (continued) - ISR

Inside the interrupt service routine, we only need to perform a single call to the `TouchDetectPosition()` function and clear the interrupt flag.

```
#define TICK_PERIOD( ms)  (GetPeripheralClock() * (ms)) / 8000

void TickInit( unsigned period_ms)
{
    // Initialize Timer3
    TMR3 = 0;
    PR3 = TICK_PERIOD( period_ms);
    T3CONbits.TCKPS = 1;      // Set prescale to 1:8
    IFS0bits.T3IF = 0;        //Clear flag
    IEC0bits.T3IE = 1;        //Enable interrupt
    T3CONbits.TON = 1;        //Run timer
}
```

Listing 4.7 - Touch Demo (continued) – TickInit() function

Touch Input - 105

The `TickInit()` function takes care of setting up the timer for the desired sampling frequency and enabling the corresponding interrupt service.

Finally we see the `main()` function. It takes care of all initializations, including: the LCD terminal emulator, the touch timer and eventually the touch state machine itself.

```c
int main( void )
{
    char s[64];

    // 1. init the graphics
    LCDInit();
    DisplayBacklightOn();

    // 2. init the touch timer
    TickInit( 1);

    // 3. init touch module
    TouchHardwareInit( NULL);

    // 4. clear screen and title
    LCDClear();
    LCDPutString( "Raw Touch Demo\n");

    // 5. main loop
    while( 1 )
    {
        if (( TouchGetRawX() != -1) && ( TouchGetRawY() != -1))
        {
            sprintf( s, "\n  %d, %d", TouchGetRawX(), TouchGetRawY());
            LCDPutString( s);
        }
    } // main loop
} // main
```

Listing 4.8 - Touch Demo (continued) – main() function

Note how the generic `TouchHardwareInit()` function calls for a pointer to be passed as its one and only parameter. In reality, in the case of the resistive touchscreen, this is not used and a `NULL` pointer is expected.

In the *main loop*, we test to see if the screen is being touched. The two functions `TouchGetRawX()` and `TouchGetRawY()` provide the actual readings of the analog to digital converter count when sampling the X and Y axis respectively but they return a negative value when no point of pressure is detected on the screen.

Build the project and then load it on the Mikromedia board to test.

If we have correctly enabled the LCD terminal emulation and added the `LCD_SCROLL` symbol to the *LCDConfig.h* configuration file, pressing the screen with a finger or a stylus will produce a continuous scrolling list of coordinates.

Calibration

Notice how the X and Y readings are not expressed directly in pixels but appears to be numbers varying approximately from 100 to 800. Also the X reading appears to be somewhat opposite the normal direction of the x axis on the cartesian plane.

Let's change now the touch screen initialization section at point 3 of the `main()` function. This time we will use the `TouchInit()` function in place of `TouchHardwareInit()`. Similarly we will use `TouchGetX()` and `TouchGetY()` instead of the similar `Raw` functions to obtain a calibrated value:

```
// 3. init touch module (do not use NVM to store calibration data)
TouchInit( NULL, NULL, NULL, NULL);

LCDClear();

// 5. main loop
while( 1 )
{
    if (( TouchGetX() != -1) && ( TouchGetY() != -1))
    {
        sprintf( s, "\n %d, %d", TouchGetX(), TouchGetY());
        LCDPutString( s);
    }

} // main loop
} // main
```

Listing 4.9 - Touch demo main() function (updated)

> **NOTE**
>
> `TouchInit()` takes *four* pointers as its parameters. Their purpose will be explained in the next chapter as we will introduce support for non volatile memory storage. For the time being, in this and the following examples in this chapter we will simply pass four `NULL` pointers values.

Let's build the project and run again the demonstration application.
Now at each new run of our code, we are being asked to proceed through a four point touch screen calibration procedure!
As a consequence, later in the main loop (5.), we obtain true screen X and Y coordinates.

```
70,  74
74,  90
83, 102
90, 113
93, 122
97, 132
103, 142
107, 149
114, 156
120, 163
131, 170
146, 180
```

Figure 4.3 Calibrated touch coordinates, screen capture

Touching What

Now that we know how to get the X,Y coordinate pair of a point of pressure on the screen, we need to figure out how to connect it to a graphical element of the user interface and eventually an action.

There are a number of user interface scenarios that we can distinguish:

1. *Tap screen to acknowledge*: when we want to wait for the user to acknowledge a (full screen) message, start and/or stop an action, we can simply sit in a loop while performing a periodic call to `TouchGetX()` waiting for a positive value to be returned.

   Here is an example of a typical splash screen:

   ```
   LCDCenterString( -1, "Important Message!");
   LCDCenterString(  1, "Tap to continue");
   while( TouchGetX() < 0);
   ```

   Listing 4.10 - Tap to Acknowledge

   Let's consider also the need to wait for the touch event to terminate (release) otherwise the application could fly through the next acknowledge request. That is no more difficult to implement than a second while loop where we wait while the value returned is positive:

   ```
   while( TouchGetX() < 0);   // wait for press
   while( TouchGetX() > 0);   // wait for release
   ```

   Listing 4.11 - Tap and Release to Acknowledge

Clearly these are tight blocking loops. In a non-blocking situation (main loop) you will need to perform the polling and move through a simple state machine.

2. *Select a direction: up, down, center:* this is an equally common use of the touch input. Whether up and down are associated to increase/decrease a controlled variable or scrolling in a list of options presented, the trick is to split the screen in two (or three) sections. Once the Y coordinate is obtained for example, we can easily discriminate the position by performing a single division:

```
if ( (y = TouchGetY()) > 0)
{
    switch( y / ( GetMaxY()/3))
    {
      case 0:   // going up
         break;

      case 1:   // select or do nothing
         break;

      case 2:   // going down
         break;
    }
}
```

Listing 4.12 - Select Up, Down and Center

Just like when reading a mechanical switch though, we might want to perform a bit of de-bouncing, or at least limit the reaction speed by inserting a delay loop to allow for human reaction times. Once more, this would be a blocking approach suitable in selected cases. In a non-blocking situation, a state machine would have to be developed to keep track of each individual press and release event.

3. *Select an element on a grid*: this is equivalent to the scenario above, but generalized simultaneously for both axis and for any number of horizontal and vertical slices. In the next section we will develop a basic demonstration and, eventually before the end of the chapter, we will turn it into a more entertaining final exercise.

4. *Select an object on the screen*: this is the ultimate and most flexible of all scenarios but also the most complex one to implement. Any number of objects painted on the screen in the shape of buttons, sliders and switches, can be made to respond to touch input by performing an association between the screen coordinates of the area covered by the object and the (X,Y) coordinate pair of the touch event. Fortunately the Graphics library comes to our rescue and provides a complete framework, the Graphics Object Library or GOL, to allow us to build with relative ease complex touch enabled graphical user interfaces. We will cover this subject in great detail in Chapter 7 of this book.

On a Grid

When we assume that a user is going to interact with a touchscreen without using a stylus but only the tips of his fingers, we have to reduce considerably our expectations in terms of the achievable resolution. In truth on a 3.2" screen, once we consider the size of a typical fingertip, we are probably looking at the possibility to discriminate no more than a grid of 5x6 positions (assuming a landscape orientation).

In the following we are going to develop a small set of functions that will help us handle touch recognition as a simple grid of buttons, albeit *invisible* ones for now.

Let's begin by defining a simple structure where we will gather the information relative to a single touch event:

```
typedef struct
{
    unsigned       x: 4;       // x coordinate on the touch grid
    unsigned       y: 4;       // y coordinate on the touch grid
    unsigned   valid: 1;       // flag, indicates a valid touch event detected
    unsigned  option: 1;       // flag, can be used to capture long touch events
} code_t;
```

It is not a coincidence that the `code_t` structure is compact enough to fit into a single 16-bit word. This makes it possible to return the entire object directly from a function.

```
code_t TouchGet( void)
{
    int x, y;
    code_t r;

    r.valid = 0;

    // 1. get the latest reading
    x = TouchGetX();   y = TouchGetY();

    // 2. if one of the two is null the other is too
    if (( x < 0) || ( y < 0))
        return r;         // return if no valid code found

    // 3. identify point on grid (4x4)
    r.y = y / ( Yside); r.x = x / ( Xside) ;
    r.valid = 1;
    r.option = 0;

    return r;

} // TouchGet
```

Listing 4.13 – TouchGet encoding function

The `TouchGet()` function in listing 4.13 polls the touchscreen status and, if a pressure is detected, it returns a `code_t` structure filled with the x and y coordinates of the touch event and a flag (`valid`) to indicate success.

Note that the x and y coordinates are transformed to *grid coordinates* by dividing the touch coordinates by the *size* of the grid: `Xside` and `Yside`. These two values need to be initialized before use and the perfect place to do so would be a `setGrid()` function possibly invoked in its turn by a comprehensive `TouchGridInit()` function:

```
static int Xside=1, Yside=1;

void setGrid( int x, int y)
{
    Xside = (GetMaxX()+1)/x;   Yside = (GetMaxY()+1)/y;
}

void TouchGridInit( int x, int y)
{
    // define the grid dimensions
    setGrid( x, y);

    // init the touch timer
    TickInit( 1);

    // init the touch state machine
    TouchInit( NULL, NULL, NULL, NULL);
} // Touch Grid Init
```

Listing 4.14 – TouchGridInit() and setGrid() functions

What we need next is a simple blocking and debouncing function, very similar in its logic to a traditional mechanical button input filter.

In fact the `TouchGrid()` function shown in listing 4.15 handles the code returned by `TouchGet()` as if it was a keyboard scan code from a traditional keypad. Since the function attempts to discriminate accidental (short) touches of the screen from legitimate events (1.), it must make use of a timed delay function. Later it expects the touch event to terminate properly once more using a timed delay loop (2.)

During the pressed state, it does also keep track of the duration of the event and eventually (3.) it will add a second flag (`option`) to the code received from `TouchGet()` to indicate a touch event longer than a predefined amount of time.

```
code_t TouchGrid( void)
{   // wait for a key pressed and debounce
    int released = 0;           // released counter
    int pressed = 0;            // pressed counter
    code_t code;                // grid code
    code_t r;                   // return value

    // 1. wait for a key pressed for at least 10 loops
    while ( pressed < 10)
    {
        code = TouchGet();
        if ( code.valid > 0) pressed++;
        else   pressed = 0;

        DelayMs( 1);
    }

    // 2. wait for key released for at least 10 loops
    while ( released < 10)
    {
        code = TouchGet();
        if ( code.valid > 0)
        {
            r = code;
            released = 0;       // not released yet
            pressed++;          // still pressed, keep counting
        }
        else released++;

        DelayMs( 1);
    }

    // 3. check if a button was pushed longer than 500ms
    if ( pressed > 500)
        r.option = 1;

    // 4. return code
    return r;

} // TouchGrid
```

Listing 4.15 – TouchGrid de-bouncing function

Creating the TouchGrid module

Since the functionality offered by these functions is going to be useful in the next few project and chapters, we can take the opportunity to create a new small shared module that we will call *TouchGrid.c.*

This will require only a small banner and a few includes:

```
/*
 * File:    TouchGrid.c
 *
 * Requires: MLA 1306, LCDTerminal.c, TouchScreen.c
 */
#include "HardwareProfile.h"
#include "TouchGrid.h"
```

This will be followed by the `TouchGridInit()`, `TouchGet()` and `TouchGrid()` functions presented in the previous few pages.

A small header file, that we will call ***TouchGrid.h,*** will help expose the function prototypes and define the `code_t` structure type:

```
/*
 * File:    TouchGrid.h
 *
 * Requires: MLA 1306
 */
#ifndef _TOUCH_GRID
#define _TOUCH_GRID

#include <HardwareProfile.h>
#include <TimeDelay.h>
#include <TouchScreen.h>

typedef struct
{
    unsigned        x: 4;       // x coordinate on the touch grid
    unsigned        y: 4;       // y coordinate on the touch grid
    unsigned    valid: 1;       // flag, indicates a valid touch event detected
    unsigned   option: 1;       // flag, can be used to capture long touch events
} code_t;

/**
 * @brief   Initializes the Resistive touch interface (uses Timer3) and defines
 *          a grid to simplify touch input
 *
 * @param x  dimension of the horizontal grid (example: GetMaxX()/4)
 * @param y  dimension of the vertical grid (example: GetMaxY()/4)
 */
void TouchGridInit( int x, int y);

/**
 * @brief   Fetches current touch state machine output and packs it in code_t
 *          structure
 *
 * @return  code_t containing current status
 */
code_t TouchGet( void);

/**
 * @brief   Blocking wait for a touch and release event. Debounces and detects
 *          event duration (setting the option flag)
 *
 * @return
 */
code_t TouchGrid( void);
#endif
```

Listing 4.16 - TouchGrid.c module

Make sure to save it inside the ***uMB*** folder, next to the other touch screen support files, for easy sharing.

A Grid Demo

Let's create a new project, that we will call **4.2-Grid**, using the checklist we developed at the beginning of this chapter and making sure to include the LCD terminal emulation module, the touch screen module and most importantly the new TouchGrid module.

Borrowing from the previous example in this chapter, we can start populating the include section, the interrupt service routine and the timer initialization function:

```
/*
 * Project: 4.2-Grid
 *
 * File:    main.c
 *
 * Requires: MLA 1306
 */
#include "PICconfig.h"
#include "LCDTerminal.h"
#include "TouchScreen.h"
#include "TouchGrid.h"

#define __ISR   __attribute__((interrupt, shadow, no_auto_psv))

void __ISR _T3Interrupt( void)
{
    _T3IF = 0;
    TouchDetectPosition();
}

#define TICK_PERIOD( ms)   (GetPeripheralClock() * (ms)) / 8000

void TickInit( unsigned period_ms)
{
    // Initialize Timer3
    TMR3 = 0;
    PR3 = TICK_PERIOD( period_ms);
    T3CONbits.TCKPS = 1;        // Set prescale to 1:8
    IFS0bits.T3IF = 0;          // Clear flag
    IEC0bits.T3IE = 1;          // Enable interrupt
    T3CONbits.TON = 1;          // Run timer
}
```

Listing 4.17   Grid project, main.c (first part)

As usual, we have included instructions for the microcontroller configuration (*PICconfig.h*). Similarly we made sure to include the definitions of the LCD terminal emulation module and the touch screen interface.

In listing 4.18, the `main()` function provides the usual initialization sections for the LCD Terminal module (1.), the timer used to drive the state machine (2.) and then proceeds to launch the touch calibration procedure (3.).
Eventually a splash screen is provided in (4.). Here we apply for the first time the basic acknowledgement technique, as described in the previous section, before the screen is cleared to make room for the actual main loop, the core of the application.

```
int main( void )
{
    code_t q;
    SHORT sx = (GetMaxX()+1)/4;
    SHORT sy = (GetMaxY()+1)/4;

    // 1. init the graphics
    LCDInit();
    DisplayBacklightOn();

    // 2. init grid
    TouchGridInit( 4, 4);
    LCDClear();

    // 3. splash screen
    LCDCenterString( -1, "Grid Demo");
    LCDCenterString( +1, "Tap to start");
    while( TouchGetX() < 0);    // wait for tap
    while( TouchGetX() > 0);    // wait for release
    LCDClear();
```

Listing 4.18 Main() function initialization section

Listing 4.19, continues with the main application loop. Here we briefly draw a (dashed) grid on the screen (5.) before invoking `TouchGrid()`, waiting for a touch event (press and release) to be detected (6.).

```
    // 4. main loop
    while( 1 )
    {
        int x, y;

        // 5. draw a grid
        SetColor( LIGHTGRAY);
        SetLineType( DASHED_LINE);
        for( x=1; x<4; x++)
            Line( x*sx, 0, x*sx, GetMaxY());
        for( y=1; y<4; y++)
            Line( 0, y*sy, GetMaxX(), y*sy);

        // 6. wait for touch on the grid
        q = TouchGrid();

        // 7. remove previous images
        SetColor( LCD_BACK);
        ClearDevice();

        // 8. choose color based on how long the pressure has been applied
        SetColor( ( q.option) ? BRIGHTRED : LCD_FORE );

        // 9. position a filled tile on the grid
        y = sy * q.y;
        x = sx * q.x;
        FillBevel( x+5, y+5, x + sx-5, y + sy-5, 5);
    } // main loop
} // main
```

Listing 4.19 – Grid.c main function, main loop

We clear the screen contents in (7.) and choose a color (8.) based on the touch event duration (`option` flag):

- *Green*, will indicate normal touch events
- *Red*, will indicate touch events that were longer than a predetermined threshold

Eventually (9.) we paint a little tile (using a Filled Bevel primitive) at the grid coordinates obtained.

Let's build the project and have fun testing the application!
Let's see what happens when we apply short and long touches, moving a finger on the screen or using multiple fingers.

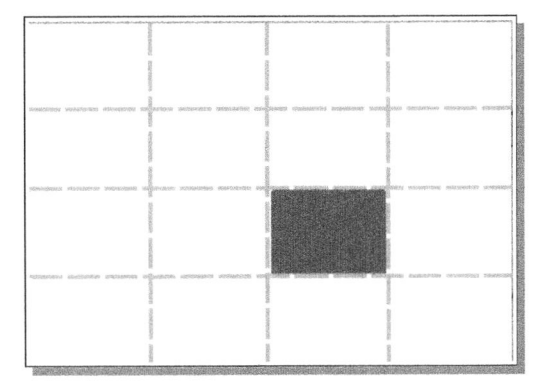

Figure 4.4 - Touch on a 4x4 grid, screen capture

Summary

In this chapter we have taken a first brief look at touch screen sensing. We have identified the support modules available inside the MLA Board Support Package and configured them for use with Mikromedia boards. For maximum simplicity we have reduced the input to a small matrix in order to satisfy a few basic use cases. We will keep expanding on the subject in Chapter 5 and 7 as we will learn to access non-volatile storage resources and later we will approach the Graphics Object Layer.

Tips & Tricks

Skipping Calibration

If you found annoying having to repeatedly calibrate the touch screen while working the previous examples, here are some simple instructions that will allow you to skip the step altogether.

1. In the example code, replace *TouchInit()* with :

```
TouchHardwareInit( NULL);
TouchCalculateCalPoints();
```

2. Edit the default values assigned in *TouchScreenResistive.h*:

```
// Default calibration points
#define TOUCHCAL_ULX 793
#define TOUCHCAL_ULY 145
#define TOUCHCAL_URX 147
#define TOUCHCAL_URY 186
#define TOUCHCAL_LLX 785
#define TOUCHCAL_LLY 793
#define TOUCHCAL_LRX 154
#define TOUCHCAL_LRY 814
```

The touchscreen should continue to work with any Mikromedia board with sufficient precision for a finger activated user interface.

Suggested Reading

- Di Jasio - **"Programming 16-bit Microcontrollers in C"** - Newnes Elsevier

    Chapter 5 – Interrupts

    Chapter 10 – It's an Analog World

- AN1478 - mTouch Sensing Solution Acquisition Methods Capacitive Voltage Divider

- AN1334 - Techniques for Robust Touch Sensing Design

- AN1325 - mTouch™ Metal Over Cap Technology

- AN1317 - mTouch Conducted Noise Immunity Techniques for the CTMU

- AN1250 - Microchip CTMU for Capacitive Touch Applications

Online Resources

- http://www.flyingpic24.com

    You can find here links to all the resources needed for this and my previous books, including code repositories, errata and updates

- http://www.microchip.com/mTouch – Touch and Input Sensing Solutions design center

- http://www.microchip.com/GestIC – 3D Gesture detection solutions

Exercises

- Modify the previous example and turn the Mikromedia board in a *fifteen tiles* game.
    This will require you to *slice* a 320x240 bitmap in a 4x4 array of tiles.
    The application will place 15 of the 16 tiles on screen. The player will touch a tile to move it to the adjacent empty space, scrambling the image at first and later to restore the correct order and reconstruct the original image.

Solutions

1. Use the GIMP Slicing Filter to split automatically the bitmap in 16 image files. (see http://docs.gimp.org/en/python-fu-slice.html – a GIMP Image Slicing Filter)
2. Use the GRC tool to pack 15 of the images in a resource file called *bitmaps.h*
3. Create a new project called *4-15_Tiles* and a new ***main.c*** file:

```c
/*
 * Project: 4-15_Tiles
 * File:    main.c
 *
 * Requires: MLA 1306
 */
#include "PICconfig.h"
#include "LCDTerminal.h"
#include "TouchScreen.h"
#include "TouchGrid.h"
#include "TimeDelay.h"

#include "bitmaps.h"      // load tiles from bitmaps
#include "droid36.h"

unsigned sx = (GetMaxX()+1)/4;
unsigned sy = (GetMaxY()+1)/4;

unsigned m[4][4];
const void * images[] = {
    &IMG0_0, &IMG1_0, &IMG2_0, &IMG3_0,
    &IMG0_1, &IMG1_1, &IMG2_1, &IMG3_1,
    &IMG0_2, &IMG1_2, &IMG2_2, &IMG3_2,
    &IMG0_3, &IMG1_3, &IMG2_3, NULL
};
```

Listing 4.20 - main.c , includes and 4x4 tiles matrix declaration

4. Include the Timer3 ISR, `TickInit()`, `TouchGet()` and `TouchGrid()` functions from the project *4-Grid*

5. Define two new functions to paint tiles and swap tiles:

```
void paintTile( unsigned tx, unsigned ty)
{
    unsigned q = m[tx][ty];  // identify the tile
    unsigned y = ty * sy;    // compute vertical coord
    unsigned x = tx * sx;    // compute horiz coord
    unsigned r = 8;          // tile rounding radius
    char s[3];

    if ( q < 15)
    {
            PutImage( x, y, (void*)images[ q], IMAGE_NORMAL);
    }
    else
    {
        SetColor( LCD_BACK);
        Bar( x, y, x+sx-1, y+sy-1);
    }
} // paintTile
```

Listing 4.21 - paintTile() function

```
void swapTiles( unsigned x, unsigned y, int dx, int dy)
{
    // move the tile in the empty space
    m[x+dx][y+dy] = m[x][y];
    paintTile( x+dx, y+dy);

    m[x][y] = 15;
    paintTile( x, y);
} // swapTiles
```

Listing 4.22 - swapTile() function

6. Modify the main function to paint the initial grid after the splash screen (5.)
7. In the main loop (6.) perform a simple check to verify which of the neighboring tiles is empty (if any) to determine the direction to move.

```c
int main( void )
{
    unsigned x, y, k;
    code_t   q;

    // 1. init the graphics
    uMBInit();
    LCDInit();
    DisplayBacklightOn();

    // 2. init the grid
    TouchGridInit( 4, 4);

    // 3. splash screen
    LCDClear();
    LCDCenterString( -3, "15 Tiles Game");
    LCDCenterString( -1, "Select to start");
    q = TouchGrid();

    // 4. init the 4x4 matrix
    LCDClear();
    k = 0;
    for( y=0; y<4; y++)
        for( x=0; x<4; x++)
        {
            m[x][y]= k++;
            paintTile( x, y);
        }

    // 5. main loop
    while( 1 )
    {
        q = TouchGrid();
        x = q .x;
        y = q .y;

        // check if near the 0 tile
        if ((x>0) && ( m[x-1][y] == 15))
            swapTiles( x, y, -1, 0);
        if ((x<3) && ( m[x+1][y] == 15))
            swapTiles( x, y, +1, 0);
        if ((y>0) && ( m[x][y-1] == 15))
            swapTiles( x, y, 0, -1);
        if ((y<3) && ( m[x][y+1] == 15))
            swapTiles( x, y, 0, +1);

    } // main loop
} // main
```

Listing 4.23 - main() function

Chapter 5

Storage

In this chapter, we will explore data storage options available on the Mikromedia boards and the related support libraries found in the Microchip Libraries for Applications. These will include access to (micro) SD cards, a Serial Flash chip and their use in combination with a file system library, touch screen and the graphic library.

Accessing a File System

MDD File System is the historical name in the MLA for the library that gives us access to micro SD cards and, most importantly, allows us to read and write files in a format that is compatible with most personal computers.

The library source files can be found in the *Microchip/MDD File System* folder. I am sure that there was a reason for the "MDD" part in the name at some point in time but, to be honest, it does not really matter anymore. What is important to know is that this library distinguishes two layers:

- A lower *physical access* layer, where the connection with the media is established using one of the microcontroller peripheral modules. In the case of the Mikromedia boards this will be the SPI2 port complemented by a pair of I/Os: pin 6 of PORTA for the *Card Detect* signal and pin 9 of PORTG as the *Chip Select* signal (see Figure 5.1). This layer is provided by the *SD-SPI.c* module.

- A higher *file system* layer offers the common file access functions and hides the complexity of the actual file system implementation (FAT16 or FAT32).

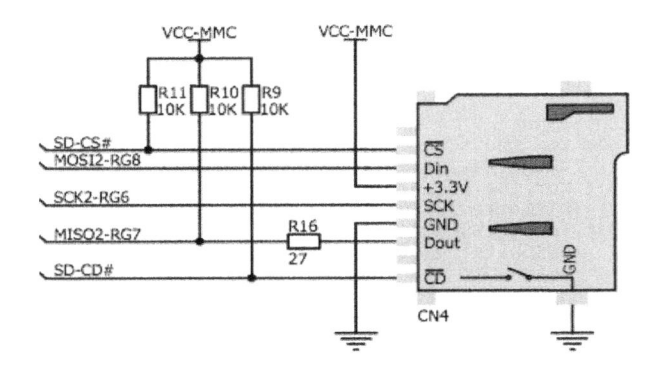

Figure 5.1 - Micro SD card slot connections

Additional physical access layer modules can be found in the same library folder to offer alternative peripherals and storage media combinations such as Compact Flash cards via a Parallel Master Port. This is just one example but the directory also has templates to help develop custom physical layer interfaces.

As with other MLA libraries, the MDD File System expects the lowest level details of the interface to be abstracted with help from an appropriate section of the *HardwareProfile.h*

```
/********************************************************************
 * IOs for the micro SD card interface
 ********************************************************************/
        #define USE_SD_INTERFACE_WITH_SPI

        // Registers for the SPI module
        #define MDD_USE_SPI_2
        // MDD SPI Configuration

        // Description: SD-SPI Chip Select Output bit
        #define SD_CS               _LATG9
        // Description: SD-SPI Chip Select TRIS bit
        #define SD_CS_TRIS          _TRISG9

        // Description: SD-SPI Card Detect Input bit
        #define SD_CD               _RA6
        // Description: SD-SPI Card Detect TRIS bit
        #define SD_CD_TRIS          _TRISA6

        // Description: SD-SPI Write Protect Check Input bit
        #define SD_WE               0
        // Description: SD-SPI Write Protect Check TRIS bit
        #define SD_WE_TRIS          _TRISA6    // not used by micro SD cards

        // Description: The main SPI control register
        #define SPICON1             SPI2CON1
        // Description: The SPI status register
        #define SPISTAT             SPI2STAT
        // Description: The SPI Buffer
        #define SPIBUF              SPI2BUF
        // Description: The receive buffer full bit in the SPI status register
        #define SPISTAT_RBF         SPI2STATbits.SPIRBF
        // Description: The bitwise define for the SPI control register (i.e. ____bits)
        #define SPICON1bits         SPI2CON1bits
        // Description: The bitwise define for the SPI status register (i.e. ____bits)
        #define SPISTATbits         SPI2STATbits
        // Description: The enable bit for the SPI module
        #define SPIENABLE           SPI2STATbits.SPIEN

        // Tris pins for SCK/SDI/SDO lines
        // Description: The TRIS bit for the SCK pin
        #define SPICLOCK            TRISGbits.TRISG6
        // Description: The TRIS bit for the SDI pin
        #define SPIIN               TRISGbits.TRISG7
        // Description: The TRIS bit for the SDO pin
        #define SPIOUT              TRISGbits.TRISG8
```

Listing 5.1 - *HardwareProfile.h* – micro SD / SPI interface section

Notice how the last part of this particular section not only selects the appropriate connections required by the Mikromedia but, also, provides a *processor model abstraction* allowing PIC32 and even PIC18 microcontrollers to be supported by the same library modules without impact to the higher layers.

Introducing FSconfig.h

Similarly a configuration file, called *FSconfig.h*, is created to select a number of file system options, mostly to the effect of optimizing the use of RAM and flash memory for the specific needs of the application. Here is a brief list of such options that we might choose to modify (enable/disable) by commenting or un-commenting individual lines in the template file provided with the library.

Make sure to make a copy of *FSconfig.h* in the project directory and store a copy in the *uMB* folder (or other folder part of the *C include dirs* path), so to become the default configuration.

Let's proceed through a quick review of the key parameters defined in *FSconfig.c*:

    #define FS_MAX_FILES_OPEN   2

The *Max Files Open* parameter specifies how many files your application will be allowed to open simultaneously. Since each file opened will use up some 500-600 bytes of RAM, you might want to keep this number small, limited to the maximum needs of your application.

    #define MEDIA_SECTOR_SIZE   512

The *Media Sector Size* is one of those parameters that you might not want to play with much until you get a much deeper understanding of the file system inner working. Let's say for now that 512 is a good value that will serve you well for most if not all but the most sophisticated of your embedded applications.

    #define ALLOW_FILESEARCH

This is the first *switch* that we encounter. If you activate (uncomment) this line, a pair of useful functions are added to your project, they are: `FindFirst()` and `FindNext()`
They will help us explore the contents of the SD cards and create user friendly selection menus, playlists and in general filtering through the potentially vast amount of data found on mass storage media.

    #define ALLOW_WRITES

Commenting this line (turning the switch off) would remove a number of library functions that allow us to *write to* and therefore *modify* the contents of the SD card. Consequently a little flash memory can be saved at the cost of a severe limitation. This might not be a bad proposition in a pure "player" application, which is not supposed to modify the contents of the media. This is also a good switch to throw when you are debugging your application and you want to be absolutely sure that a rogue program does not accidentally corrupt the SD card contents. In most all other cases, you will want it ON (uncommented).

`//#define ALLOW_FORMATS`

This is a far more powerful and dangerous switch. Unless an application MUST be able to reformat the SD card, I suggest we keep this line commented and save that space for other good uses. We'll let the user do the card formatting, when necessary, on a personal computer!

`#define ALLOW_DIRS`

While the size of most SD cards is so large nowadays that you are unlikely to ever run out of storage space in your embedded control applications, the root directory of a (FAT) file system can be limited to a maximum number of files (approx. 100) and that limit can be reached relatively quickly. Allowing support for directories and sub-directories makes life easier and can help keep things clean and orderly.

`//#define ALLOW_FSFPRINTF`

When this switch is used (uncommented), the `FSfprintfs()` function can be used to output formatted strings to a file. Small microcontrollers (PIC18) or severely (flash memory) constrained applications might not be able to accept the extra memory space required (several Kbytes) to support the complex string formatting features. In those cases we might want to do without and/or provide a more basic implementation ourself.

`#define SUPPORT_FAT32`

This is required to accept the file system formats used to support devices with 2GB or more of data storage space. Once, this used to be a luxury but, nowadays, it is becoming more of the norm. If we don't allow for FAT32 support, we are most likely going to severely limit the choice of cards that will work with our applications.

The last three defines found in the *FSconfig.h* template are actually three mutually exclusive options meant to instruct the file system library on how to handle *timestamps*. Each file written into the file system needs to have a time and a date of creation/modification recorded. The MDD File System library offers the following options so to assist in a consistent use of those fields:

`//#define USEREALTIMECLOCK`

If uncommented, the library will assume that you have enabled the Real Time Clock and Calendar (RTCC) module of the PIC24 and initialized it with the proper time and date.

```
//#define USERDEFINEDCLOCK
```

This option will expect the user to handle the values to be used for the timestamp. This could be because of a non standard method to keep track of time (deriving timing information from an external RTC for example). Before writing or modifying a file, we will have to make sure to provide an update to the relevant data structure used by the library by making a call to the function `SetClockVars()`.

```
#define INCREMENTTIMESTAMP
```

This last option is the most convenient to use when we don't give particular importance to recording an accurate timestamp to our files. It will simply provide a counter value that will increment every and each time we are creating or modifying a file.

Introducing uMedia.c

Just as I tricked you in the second chapter, I could now have you trying to use the MDD File System library right away only to have you discover that the micro SD card cannot be accessed yet. Only after some serious debugging effort you would probably discover that the SPI2 peripheral is not actually reaching the I/O pins it is supposed to!

This is because, unless we configure the Peripheral Pin Select feature of the PIC24FJ256GB110, the SPI peripheral signals are not mapped to the desired pin locations yet. This is an initialization that needs to be performed at run time and therefore cannot be included in the *PICconfig.h* file. Using the peripheral library module *pps.h*, we could quickly fix this problem, with the following few lines of code added to the top of the `main()` function:

```
// SPI2
    PPSInput( PPS_SDI2,  PPS_RP26);      // SDI2 =RP26 G7/pin 11
    PPSOutput( PPS_RP21, PPS_SCK2OUT);   // SCK2 =RP21 G6/pin 10
    PPSOutput( PPS_RP19, PPS_SDO2);      // SDO2 =RP19 G8/pin 12
```

Listing 5.2 - PPS-SPI2 configuration example

But, since we will be using the SPI peripheral consistently in the following several chapters to reach a number of other hardware modules (accelerometer, audio codec, serial flash), we should rather take the opportunity to create a shared module.

We will call this module *uMedia.c* and we will save it in the *uMB* folder from where it will be accessible to each project via the *C include dirs* path.

```
/*
 * uMedia.c
 *
 * PIC24 Mikromedia board
 */
#include "uMedia.h"
#include <pps.h>

void uMBInit( void)
{
    // 1. disable analog inputs
    AD1PCFG = 0xFFFF;    // all inputs digital

    // 2. configure PPS for PIC24 Mikromedia
    PPSUnLock;

    // 3. Configure SPI2
    PPSInput( PPS_SDI2,  PPS_RP26);       // SDI2 =RP26 G7/pin 11
    PPSOutput( PPS_RP21, PPS_SCK2OUT);    // SCK2 =RP21 G6/pin 10
    PPSOutput( PPS_RP19, PPS_SDO2);       // SDO2 =RP19 G8/pin 12

    // Done, optionally lock the PPS configuration
    // PPSLock;

} // uMBInit
```

Listing 5.3 – uMedia.c, Initialization Function

We can also take the opportunity to add the `TickInit()` function and the Timer3 ISR function, as defined in Chapter 4, and place them inside the *uMedia.c* module.

Let's add the function prototypes to the ***uMedia.h*** header file:

```
/*
** uMedia.h
**
**    Mikromedia configuration and basic peripherals access
**
*/

#ifndef _MIKROE_uMB
#define _MIKROE_uMB

#include <xc.h>
#include "HardwareProfile.h"           // need to know clock freq.

#include "Touchscreen.h"
#include "Graphics/Graphics.h"

// function prototypes and macros
void uMBInit( void);             // init standard hw configuration
void TickInit( unsigned);        // init Timer3 as the touch timer

#endif
```

Listing 5.4 - uMedia.h

NOTE FOR PIC32 USERS

Users of the PIC32 Mikromedia board models will find that the PPS setting is NOT required on their boards, but will find the new uMedia module helpful to initialize the clock system, cache memory support and any number of additional features that are specific to the PIC32 processor to achieve maximum performance.

Reading Text Files

With this new tool under our belt, we are ready to try a first experiment. We are going to test our ability to initialize the SD card, open a specific file, and list its contents (at least one buffer full) on the display used as an alphanumeric terminal.

This is a good exercise that gives us reason to explore the most basic functions offered by the MDD File System library. If you have any experience with C language standard file I/O libraries, you will find the MDD library quite familiar. In fact, in most cases, you will only need to add the *FS-* prefix to the function and structure names.

Preparation

Let's create the new project and populate it with the basic elements of the Graphics library, LCD terminal emulation, touch screen and some of the MDD File System new components. Here are the simple steps:

1. Use the **New Project Wizard** to create a new project inside the working directory, let's call it: "**5.1-Storage_Text**"

2. Add a new logical folder to the Source Files folder, called **MLA**

3. Add the following items to the folder:

- *IL9341.c (or HX8347.c)*, the display controller driver (found in *Graphics/Drivers*) depending on the display model

    - *Primitive.c* , the primitive layer of the graphics library (found in *Graphics*)

    - *TimeDelay.c*, a few basic timing functions used in the driver (found in *Common*)

    - *FSIO.c*, the FAT file system support library (found in *MDD File System*)

    - *SD-SPI.c*, the low level SPI interface layer of the file system (found in *MDD File System*)

4. Add a new logical folder to the Source Files folder, let's call it: **uMB**

5. Add the following items found in the corresponding folder:
   - *LCDTerminal.c*, the terminal emulation library
   - *LCDTerminalFont.c*, the terminal emulation mono-spaced font
   - *TouchScreen.c,* support for generic touchscreen input
   - *TouchScreenResistive.c,* specific support functions for resistive touchscreens
   - *TouchGrid.c,* the simplified touch debouncing and 4x4 grid reduction we developed in the previous chapter
   - *uMedia.c,* provides the `uMBinit()` function to initialize pins and ports
6. Configure the compiler *C include dirs* path to contain:
   - **. (dot)**, the current project directory for MLA to reach our configuration files
   - *../Microchip/Include*, for our source files to reach inside the MLA
   - *../uMB*, for project source files and the MLA to access the Hardware Profile and other resources shared among this and future projects in this book and specific to the Mikromedia board.
7. Create a new main file that we will call *main.c* using the **New File wizard** and the embedded template (or your own customized version)

Repetita Juvant or does it?

If you followed along the chapters of this book in a linear fashion, I am sure that at this point you will have noticed how we have gradually increased the number of source files that we include in every and each project. In the above checklist we have just added two more (*FSIO.c* and *SD-SPI.c*) from the *MDD File System* library on top of the now familiar lot. You can rest assured that the list will keep growing for a while, as we will cover all the remaining peripheral modules present on the Mikromedia board and a number of optional modules too.

If this sounds like a lot of repetition, keep in mind that the way we proceed in our quest in this book is not representative of a typical day at work. More often than not, we will be setting up a project every once in a while and then work on it for weeks in a row.

On the other hand there are several tricks that can be used to make life easier once familiar with the basic MLA structure and MPLAB® X.

For example it is possible to *copy and rename* entire projects, which means we can create *project templates*, ready to use and customize.

It is also possible, although a bit more challenging, to pre-assemble any number of modules in a custom *library archive (.a)* file that could be specific for a given hardware platform (such as the PIC24 Mikromedia). Multiple projects that use the same hardware platform could then be able to simply link it in by adding the archive file to the *Libraries* logical folder in the MPLAB X Project window.

Back on track

Let's proceed by reviewing step by step the code in *main.c* (Listing 5.5):

```
/*
 * Project: 5.1-Storage_Text
 * File:    main.c
 *
 * Requires: MLA 1306
 */
#include "PICconfig.h"          // device configuration

// 1. include libraries
#include "TimeDelay.h"
#include "LCDTerminal.h"
#include "TouchGrid.h"

#include "MDD File System/FSIO.h"

char data[ 400];                // a buffer of arbitrary lenght

int main( void )
{
    FSFILE *fp;                 // pointer to a file structure
    unsigned length;
    char *p;

    // 2. initializations
    uMBInit();                  // init pins and ports
    LCDInit();                  // inits terminal emulation
    DisplayBacklightOn();
    TouchGridInit( 3, 3);       // defines a 3x3 grid

    // 3. splash screen
    LCDClear();
    LCDCenterString( -1, "TEXT demo");
    LCDCenterString( +1, "tap to start");
    TouchGrid();
    LCDClear();
```

Listing 5.5 – Storage_Text, *main.c*, first part

1. In addition to the *TimeDelay.h, LCDTerminal.h and TouchGrid.h* files this time we are adding: *MDD File System/FSIO.h*
2. Next we proceed to initialize the Mikromedia board I/Os by calling the function uMBInit(), we initialize the Terminal emulation with LCDInit(), we turn on the display backlight and eventually initialize the touchscreen grid module.

3. A few lines later, we present a basic splash screen and we wait for the user touch to start the demo.
4. At this point (Listing 5.6), we enter the main loop and after clearing the screen, we proceed to try and initialize the SD card with `FSInit()`.

   Note how this function can fail for a number of reasons. It can be simply because the SD card is not inserted or it has not completed its power up cycle yet. But it could also be that the format used to store data on the card is not compatible (not FAT16 or FAT32). In any case, the best thing we can do is prompt the user and repeat the test every so often (100ms) until a suitable card is inserted.

```
// Main Loop
while( 1)
{
    LCDClear();

    // 4. init file system, wait for SD card to be inserted
    while  ( FSInit() != TRUE)
    {
        LCDCenterString( 0, "Insert SD Card");
        DelayMs( 100);
    }

    // 5. try to open a file
    if ( (fp = FSfopen( "README.TXT", "r")) == NULL)
    {
        LCDPutString( "\n File Not Found");
    }
    else    // file found
    {
        // 6. output the content of the first buffer read
        length = FSfread( data, 1, sizeof(data), fp);
        p = &data[0];
        while( length-- > 0 )
        {
            LCDPutChar( *p++);
        }

        // 7. print "..." if the file continues
        if ( !FSfeof( fp))
            LCDPutString( "...");

        FSfclose( fp);

    }   // else

    // 8. prompt to continue
    LCDPutString( "\n Tap to continue");
    TouchGrid();

}   // main loop
}   // main
```

Listing 5.6 – Storage Text, *main.c*, (continued)

5. We try to open a file for reading, looking specifically for the filename: *README.TXT* using the `FSfopen()` function. If successful, it will return a pointer to a `FSFILE` structure or `NULL` in the event the file could not be found.

6. We attempt to read an entire buffer, represented by the array `data`, with the `FSfread()` function which returns the number of characters actually read and we proceed to print its contents on the screen.
7. We check if have not reached the end of file already with `FSfeof()`, in which case we simply print a three dots sequence "..." and we close the file with `FSfclose()`.
8. The whole sequence contained in the main loop can then be repeated after we prompt the user and wait for a new touch event.

Building the project and programming the Mikromedia board, we will quickly be able to prove that we can visualize the contents of any text file. You must ensure that the SD card is formatted with the right kind of file system. Windows users will find that *NTFS* is not an acceptable option, as well as Linux and OS X users will find that *Ext3* and other more advanced formats are not accepted. Luckily all personal computers users nowadays will be able to use the common denominator represented by FAT16 and/or FAT32.

While our example assumes an ASCII text file for simplicity, it does not have to be. Any binary content could be read with the exact same sequence of operations and function calls, we simply would not be able to "print" it on the terminal, but we could perhaps provide a "hex dump" for demonstration purposes.

Selecting a File from a List

Clearly this first example project offered just a glimpse into the possibilities offered by the MDD library. Without increasing the complexity by much, we can now quickly add a bit more user friendliness to the application by allowing the user to choose a file name from a list of files found on the micro SD card (root) directory. This gives us the perfect opportunity to explore two new functions: `FindFirst()` and `FindNext()` that are enabled by the presence of the `ALLOW_FILESEARCH` symbol in *FSconfig.h*.

The `FindFirst()` function is designed to inspect the current directory (the root by default) and check each file name found against a given search-string using a simplified sort of regular expression. The pattern matching scheme is based on the use of two special characters "*", "?" and will be familiar to all readers with a minimum of experience in the use of a typical OS command line.

As an additional selection parameter, the function can be told to inspect only proper file names vs. folder names and/or to match other *file attributes*. Since many files could provide a positive match, `FindFirst()` will provide pointers only to the first such file found and will place other information related to the search in a special `SearchRec` structure that is passed as a pointer. Should we desire to continue our search until a complete list of matching files is obtained, we can then call the corresponding `FindNext()` function repeatedly. Make sure to pass along a pointer to the `SearchRec` structure, initialized by `FindFirst()` during the first call, it will be updated by `FindNext()` at each subsequent step.

The following snippet of code (Listing 5.7) should illustrate a simple usage of the technique described:

```c
char list[ N_FILES][16];
int n = 0;
SearchRec sr;

...
    if ( !FindFirst( "*.TXT", ATTR_READ_ONLY | ATTR_ARCHIVE, &sr))
    {
        do{
            // while there are files matching
            // copy the file name
            strncpy( list[ n++], sr.filename, 16);

            // check if list full
            if ( n==N_FILES)
                break;
        } while ( !FindNext( &sr));
    }
```

Listing 5.7 – Obtaining a list of file names matching a given extension

Here a small array of strings (`list`) is filled with items matching a given string *"*.TXT"* which results in a selection by filename extension. Notice how the inner loop can be terminated both by reaching the maximum number of items the array can hold (`N_FILES`) or when no more filenames can be found by `FindNext()`.

Menus

Presenting a basic list of items, a menu, on the LCD display of the Mikromedia board and accepting touch user inputs, can be achieved in very few lines of code thanks to the LCD terminal module and the TouchGrid module we developed in the previous chapter.

Listing 5.8 illustrates a most rudimentary implementation of such a menu selection function that can be immediately used to augment our first application and allow the user to pick a text file name from the list of available ones.

The `Menu()` function will simply take a generic array of strings of given length and attempt to display it (centered) on the LCD terminal using a simple color inversion (background/foreground) technique to highlight the item currently selected.

```c
int Menu( char list[][16], int items)
{
    int i, j, n;
    code_t c;
    // 2.1 init cursor position, first item selected
    n = 0;

    // display item list and wait for user input
    while ( 1)
    {
        // 2.2 draw menu list of items
        for(i=0; i< items; i++)
        {
            // position each item on the menu centered horizontally and vertically
            LCDSetXY( (_MAX_X-14)/2, ( _MAX_Y - items)/2 + i);

            // temporary color scheme
            if (i != n)
            {   // non selected items are displayed in Yellow over Blue
                LCDSetBackground( BLUE);
                LCDSetColor( YELLOW);
            }
            else
            {   // the current selected item is displayed in blue over yellow
                LCDSetBackground( YELLOW);
                LCDSetColor( BLUE);
            }

            // each item must be padded to clear the entire line
            LCDPut(' ');
            LCDPutString( list[ i]);
            for(j=0;j<(14-strlen( list[i])); j++)
                LCDPut(' ');
        } // for

        // restore default colors
        LCDSetBackground( LCD_BACK);
        LCDSetColor( LCD_FORE);
```

Listing 5.8 - Basic Menu function (first part)

Notice that in its simplicity this code makes some important assumptions (limitations), among which:

- The number of items in the list is supposed to fit entirely in the number of lines available on the screen (no windowed scrolling is provided if the list exceeds the space available on screen)

- The width of the items is fixed to 16 characters, hence no long file names

- No check is performed here to ensure that the items do contain actual valid file names

```
        // 2.3 wait for a touch input event
        c = TouchGrid();
        if ( c.x == 2)          // right side of the screen -> select
        {
            return n+1;
        }
        else if ( c.y == 0)     // top of the screen -> move cursor up
        {
            if (n > 0)   n--;
        }
        else if ( c.y == 2)     // bottom of the screen -> move cursor down
        {
            if ( n < items-1) n++;
        }
    } // while
} // Menu
```

Listing 5.9 – Basic Menu function (continued)

The touch grid input mechanism is designed to interpret a touch event on the upper part of the screen as a request to move the selection up and respectively down when the bottom part of the screen is touched. A touch on the right side of the screen will exit the function returning the index of the currently selected item.

Forming the list of items with the technique illustrated previously in Listing 5.7, we can now write the new `SelectFile()` function.

The `SelectFile()` function (Listing 5.10) will take care of the entire process, from prompting the user to insert a card and initializing the file system, to forming the list of suitable file names and offering the user to choose one of them invoking the *menu()* function of Listing 5.8/9. Eventually it will return with the selected filename string.

```c
void SelectFile( char *fn, char *ext)
{ // fn     pointer to selected filename
  // ext    "*.TXT" selection criteria

    char list[ N_FILES][16];
    int n = 0;
    SearchRec sr;

    while( 1)
    {
        // 1. ensure the file system is initialized, card inserted
        while( !FSInit())
        {
            LCDCenterString( 0, "Insert Card");
            DelayMs( 100);
        }
        LCDClear();

        // 2. search for ".ext" files and put them in list
        if ( !FindFirst( ext, ATTR_READ_ONLY | ATTR_ARCHIVE, &sr))
        {
            do{
                // while there are files matching
                // copy the file name
                strncpy( list[ n++], sr.filename, 16);

                // check if list full
                if ( n==N_FILES)
                    break;
            } while ( !FindNext( &sr));
        }

        // 3. n = listTYPE( list, N_FILES, ext);
        if ( n > 0)
        {
            // found at least one file
            n = Menu( list, n);
        }

        // 4. if no file found or none selected
        if ( n == 0)
        {
            //report error and allow to swap card
            LCDCenterString( 0, "File Not Found");
            TouchGrid();
            LCDClear();
        }

        else // 5. valid file selected
        {
            //form the chosen filename
            strncpy( fn, list[(n-1)], 16);
            return;
        }
    }// while

} // select file
```

Listing 5.10 – The SelectFile() function

We can now add the two functions `Menu()` and `SelectFile()` at the top of the `main()` function to obtain a new and improved application (project) that we could simply call: *5.2-Storage_Menu*.

Let's build the project and run the application on the Mikromedia board to test the new file selection functionality.

Figure 5.2- Screen capture of the Storage_Menu project

Now that we know how to select and read generic text files from an SD card, it is certainly interesting to see how we can expand the horizon a bit and take on a different kind of data that is certain to require large amounts of space: back to graphics!

Reading Image Files

As we have seen in Chapter 3, graphic resources such as fonts and bitmaps can quickly grow to fill the flash memory of a typical microcontroller. A micro SD card, even one of the smallest and lowest cost available today, would be capable of storing tens of thousands of images even without considering available image compression options.

> **NOTE**
> A quick back of the envelope calculation tells us that a 2GByte micro SD card (retailing for approximately five US dollars as of this writing) would be able to store more than 12,000 full screen QVGA uncompressed images (~150Kbytes each)!

But if extracting images directly from an SD card to the screen seems an attractive option, we have to consider a few of the possible drawbacks, among which:

- Reduced speed, since the image data is to be extracted from the SD card via a serial interface (SPI), there is an inherent bottleneck in the system. This is in addition to the complexity of the file system used and the required intermediate indexing and buffering steps. Fortunately these are handled for us, transparently for the most part, by the MDD File System library.

- Increased RAM usage, even when no additional buffering is added, the file system library is using up at least a couple of "sector" buffers (512-bytes large).

In truth, there is an entirely different kind of consideration that must be added to any embedded control design each time a "connector" is added to the system: it's *mechanical!* Vibrations could dislodge the card in the slot, as well as humidity and other environmental contaminants could eventually compromise the contacts (in such cases the reader might want to investigate the use of eMMC technology instead).

Assuming these considerations have been taken into account, we will explore now another module of the Microchip Libraries for Applications: the *Image Decoders*.

Image Decoders

Let's create a new project that we will call **5.2-Storage_Slides,** copying the entire structure and contents of the previous *Storage_Menu* project but adding a few additional library source files that can be found in the *Microchip/Image Decoders* folder:

- *ImageDecoder.c*, this will bring in the generic image decoding functions

- *BmpDecoder.c*, this will handle the (Windows 16-bit) *Bitmap* file format decoding specifics
- *JpgDecoder.c*, this will handle the *JPEG* file format decoding specifics
- *Jidctint.c*, this will cover the math intensive portion of the JPEG decoding algorithm
- *GifDecoder.c*, this will handle the *GIF* file format decoding specifics

The aim of this demonstration project will be that of displaying a number of images (slides) from various files and formats read from the root directory of an SD card and selected using the simple menu mechanism developed in the previous pages.

Optionally, this application could be extended at a later time to be run as a self timed slideshow!

Creating LCDmenu

To this end we will re-purpose a portion of the previous demo project main module and turn it into a shared module by separating the two functions `Menu()` and `SelectFile()` in a new file called ***LCDmenu.c***.

We will also create a header file for it: ***LCDmenu.h*** as shown in Listing 5.11.

```
/*
 * File:    LCDmenu.h
 *
 */
#include "TimeDelay.h"
#include "LCDTerminal.h"
#include "TouchGrid.h"
#include "MDD File System/FSIO.h"

#define N_ITEMS    12      // maximum length of selection list

/**
 */
int Menu( char list[][16], int items);

/**
 */
void SelectFile( char *fn, char *ext);
```

Listing 5.11 – *LCDmenu.h* header file

We will move the two new files to the *uMB* folder to make them available to all future projects.

We will include them in our new project (5.2-Storage_Slides) Source Files list to link them to our new application with the rest of the LCD terminal and touch screen support modules.

In the *main.c* file we will insert two additional include directives:

```
#include "LCDmenu.h"
#include "Image Decoders/ImageDecoder.h"
```

The *Image Decoders* library acts as a sort of a bridge between the MDD File System library and the Graphics library and can be extended to include additional interfaces so that images can be "streamed in" from a multitude of sources.

Since the act of decoding some of the most advanced file formats (JPEG in particular) can be quite processor intensive, requiring occasionally up to a few seconds of continuous work on a 16-bit microcontroller for a large full screen image, the library makes provision for a call back mechanism, so that it is possible to avoid blocking completely the entire application and to maintain a minimum level of responsiveness.
The desired callback function can be effectively inserted in the middle of the image decoding loop by passing a pointer to the `ImageLoopCallbackRegister()` function.

Should then a long decoding operation require to be aborted for any reason, it is possible to set an appropriate flag (from inside the callback function) using the `ImageAbort()` macro. Other than that, there are only two basic functions that need to be mastered:

- `ImageDecoderInit()`, to be called only once before first use
- `ImageDecode()`, or the simplified `ImageFullScreenDecode()` which will perform the actual decoding and painting, eventually centering the resulting image to use the entire screen.

The `ImageDecode()` function requires the chosen file to be already opened (for reading) before it is called. The file pointer is passed as a *first parameter* while the *second parameter* specifies the expected file format.
Note that the format is not automatically recognized. Although it could be likely deduced from the file name and extension, this task is left to the calling application.
There are only three types of file formats known to the library as of this writing:

- IMG_BMP
- IMG_JPEG
- IMG_GIF

A *third* and *fourth parameter* are designed to allow alternate input and output methods (in place of file read and graphics display output) and will be simply passed as NULL pointers in most standard use cases.

The PutSlide() function, presented in Listing 5.12, does offer an actual example of use of ImageDecode() adding the automatic recognition of the file type by the given filename extension:

```
int PutSlide( char *filename)
{
    char *ext;
    IMG_FILE *pImageFile;
    IMG_FILE_FORMAT fmt;

    // find the file extension
    ext = filename;
    while( *ext)
    {
        if ( *ext++ == '.')
            break;
    }
    // depending on file extension deduce image format
    if (( strcmp( ext, "JPG") == 0) || ( strcmp( ext, "jpg") == 0))
        fmt = IMG_JPEG;
    else if (( strcmp( ext, "GIF") == 0) || ( strcmp( ext, "gif") == 0))
        fmt = IMG_GIF;
    else if (( strcmp( ext, "BMP") == 0) || ( strcmp( ext, "bmp") == 0))
        fmt = IMG_BMP;
    else return -2;      // slide format unknown

    // decode selected image
    pImageFile = IMG_FOPEN( filename, "r");
    if(pImageFile == NULL)
    {
        return -1;  // error: file not found
    }
    else
    {
        ImageFullScreenDecode( pImageFile, fmt, NULL, NULL);
        IMG_FCLOSE(pImageFile);
    }

    // return success
    return( 0);
} // PutSlide
```

Listing 5.12 – PutSlide() function

Configuring the Image Decoder

As most other library modules, even the image decoder has his own configuration file: *ImageDecoderConfig.h*.

Just as in all previous cases, we will copy the provided template directly into our project directory and/or in our shared modules folder (*uMB*) and from there we will make the required customizations:

```
/************* User configuration start *************/

/* Comment out the image formats which are not required */
#define IMG_SUPPORT_BMP
#define IMG_SUPPORT_JPEG
#define IMG_SUPPORT_GIF

/* Comment out if output has to be given through a callback function */

/* If defined, the code is optimized to use only the graphics driver, only 16-bit-
565-color-format is supported */
#define IMG_USE_ONLY_565_GRAPHICS_DRIVER_FOR_OUTPUT

/* If defined, the code is optimized to use only the MDD file system */
#define IMG_USE_ONLY_MDD_FILE_SYSTEM_FOR_INPUT

/* If defined, a loop callback function is called in every decoding loop so that
application can do maintainance activities such as getting data, updating display,
etc... */
//#define IMG_SUPPORT_IMAGE_DECODER_LOOP_CALLBACK

/************* User configuration end *************/
```

Listing 5.13 – ImageDecoderConfig.h

Listing 5.13 illustrates a typical configuration. Commenting or uncommenting individual file formats, when not used, will help reduce the application code size – JPEG being the largest consumer.

Commenting the `USE_ONLY_MDD_FILE_SYSTEM_FOR_INPUT` can be required when the application needs to invoke the library to decode images incoming from other sources. In such a case a structure (`IMG_FILE_SYSTEM_API`) containing pointers to alternate functions to be used to *open, read, seek, close* image streams should be defined and filled accordingly. Note how a pointer to this structure can then be provided as the third parameter to the `ImageDecode()` function.

Uncommenting `IMG_SUPPORT_DECODER_LOOP_CALLBACK` adds the calls to the callback registered function inside the deepest loop of the decoding process.

Displaying Slides

With help from the `PutSlide()` function and all the Image Decoding library modules included in the project Source Files list, we are ready to complete the application by writing our new `main()` function as in Listing 5.14.

```
int main( void )
{
    FSFILE *fp;
    unsigned length;
    char *p, filename[32];

    // 1. initializations
    uMBInit();                     // init pins and ports
    LCDInit();                     // inits terminal emulation
    DisplayBacklightOn();
    TouchGridInit( 3, 3);          // defines a 3x3 grid
    ImageDecoderInit();            // init the image decoding lib

    // 2. splash screen
    LCDClear();
    LCDCenterString( -1, "Slides demo");
    LCDCenterString( +1, "tap to start");
    TouchGrid();

    // 3. Main Loop
    while( 1 )
    {
        LCDClear();

        // 4. try to open a file
        SelectFile( filename, "*.*");
        LCDClear();

        // 5. decode and place image centered full screen
        if ( PutSlide( filename))
        {
            LCDCenterString( 0, "Unable to Display");
        }

        // 6. prompt to continue
        TouchGrid();
    }   // main loop
}   // main
```

Listing 5.14 - Storage Slides, main() function

Notice how this time we invoked the `SelectFile()` function using the "*.*" notation, therefore allowing the selection menu function to list all files available in the currently selected (root by default) directory of the SD card. In this way all files present will show up in the menu, including some that might not necessarily represent images. It will be up to the `PutSlides()` function to help us further discriminate later.

Make sure to copy some suitable images on the SD card – use the GIMP application to ensure that the maximum image size is compatible with the QVGA screen size: 320x240 pixels.

Let's build the project and load it on a Mikromedia board. Select and visualize a few files of different formats to test our newly acquired decoding capabilities!

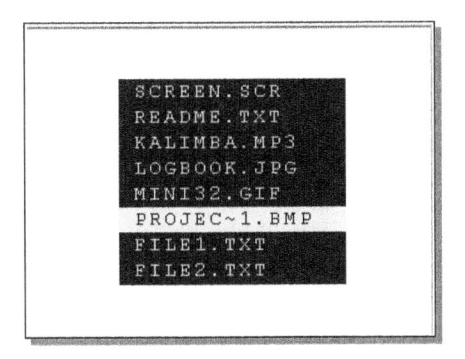

Figure 5.3 – Slides application, screen capture

Figure 5.4 – ProjectPilot.BMP slide

Choosing the Right File Format for your Images

Each image file format has its pros and cons, it is important to understand them to make the right choice for the application:
- *GIF*, an excellent format for all images that contain large section of uniform colors. It is in those cases that it achieves the best compression and speed.
  While in the past there were concerns on the proprietary nature of the algorithm used, as of this writing all existing legal constraints have been lifted.
  Note: the current library implementation is not supporting some features that are otherwise quite popular in personal computer applications such as transparency and animation!

- *BMP*, or more precisely Windows 16-bit BMP, as we should refer to it in order to distinguish it from the (too) many variants of this format in common use. It is the lightest of all three formats as it requires the least amount of resources for decoding and since there is no compression/decompression performed, speed is fully determined by the SD card data transfer (and SPI port) limitations. Of course, the disadvantage of the BMP format is that file sizes can get prohibitively large.
- *JPEG*, the preferred format to be used with photographic images and in general images with high color content. Contrary to the previous two formats, it uses a *lossy* compression algorithm, which makes it less suitable for images that contain text or high contrast edges in general. It is also the slowest format to decode by far.

You might note that support for other popular file formats is missing, most notably PNG. For many years the PNG format has been gaining popularity while replacing both GIF (to avoid the legal entanglements of the past) and the BMP complex legacy and excessive fragmentation.

Fortunately the structure of the Image Decoders library is quite modular and it is possible to add new decoders as necessary. Perhaps an interesting challenge you might want to take up?

Serial Flash

Mikromedia boards offer another useful storage facility for our applications: an 8 Megabit Serial Flash device.

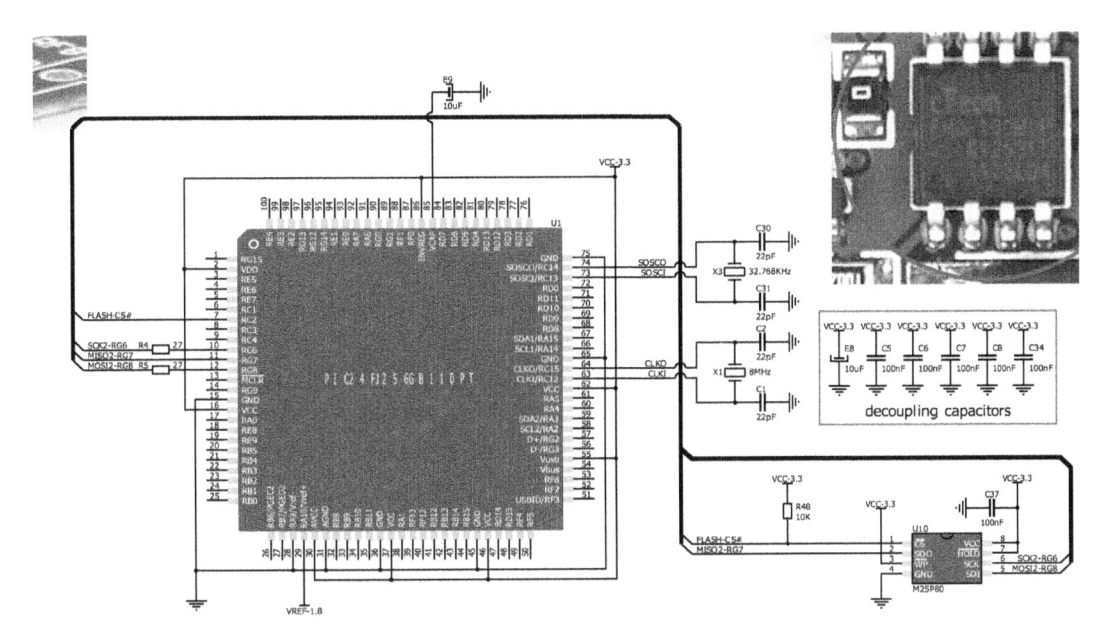

Figure 5.5 – M25P80 Serial Flash device interface

This additional 1 MegaByte of storage can be quite handy in those cases where we don't want to depend on a micro SD card – see previous consideration regarding mechanical reliability and other environmental concerns.

As we can see from the schematic in Figure 5.5 (extracted from the Mikromedia board user manual), the interface to the serial flash device is based on the same SPI2 interface used also for the micro SD card.

You will be pleased to know that the specific peripheral settings (clock polarity, edge and bit rate) are compatible among the two. In fact the only distinction between the two devices consists in the *Chip Select* line that in the case of the Serial Flash is controlled by *PORTC* pin 2.

The serial flash device used on the Mikromedia board is an *M25P80* and it belongs to a large family of devices that share a somewhat standardized SPI interface similar to that used by some of the most common serial EEPROMs.

> **NOTE**
>
> The key differences between serial EEPROM and Serial Flash devices are:
>
> 1. The size of the erase blocks, EEPROMs can be typically erased byte by byte while flash devices can be erased only in relatively large blocks
>
> 2. The endurance, that is the number of erase/write cycles that can be performed over the life of the device without incurring in data loss; EEPROMs can count millions of cycles whereas flash devices typically top in the ten of thousands
>
> 3. The sheer size of the memory array, EEPROMs top at a few hundred kilobytes where Serial Flash devices offer megabytes.

The M25P80 memory can be programmed byte by byte or in blocks of up to 256 bytes at a time (*write-pages*). But it can be erased only as 16 individual blocks (or *erase-sectors*), each containing 64 Kbytes. The entire memory can be erased at once (*bulk erase*), or it can be erased one sector at a time (*sector-erase*).

Extreme care must be taken when updating only a section of the serial flash memory array as erase and write pages are not the same size and significant buffering (up to 64K bytes at a time) would be required to perform a partial update!

Library support is currently (as of MLA revision 1306) available through the *m25p80.c* file found in the MLA *Board Support Package* folder. This, in turn, requires the *drv_spi.c* module found in the same folder, which provides the SPI low-level routines.

Visually inspecting the *m25p80.c* source file, we soon realize that the functions offered are carrying names prefixed by "SST25-" a string referring to a similar, but slightly incompatible series of serial flash devices. However, these are the function names used at the application level to issue the basic erase, read and write commands necessary.

In essence, what is happening here is that the M25P80 is treated as a "variant" of the SST25 family of devices, which is taken as *synonym* for serial flash.

> **NOTE**
>
> The *drv_spi.c* module will reveal in its turn an original design that appears to support multiple SPI peripherals using a novel indexed scheme. This is perhaps a sign of things to come (future revisions of the MLA will extend the use of this technique.

A Serial Flash Demo Project

In as few lines of code as possible we will now demonstrate how to copy the entire contents of a file (in this case simply a text file) into the serial flash memory device. From there we will read it back and display it onto the LCD terminal for verification.

As in the previous examples, we will allow the user to select the file name to be copied from those available in the micro SD card root directory using the *LCDMenu.c* module.

Preparation

As we prepare one last project, that we will call **Storage_Flash,** using the previous *Storage_Menu* project as a template, we can add the two new modules ***drv_spi.c*** and ***m25p80.c*** (copied in the *uMB* folder) to the Source Files list inside the new project window.

HardwareProfile.h Serial Flash section

Once more there is a need for a small section to be added to the *HardwareProfile.h* file in order to support the serial flash device and configure it for use with the Mikromedia board. Here is the related section that we can append to it right away:

```
/******************************************************************
 * IOs for Serial Flash
 ******************************************************************/
#define USE_M25P80
#define SPI_CHANNEL_2_ENABLE
#define SST25_CS_TRIS       _TRISC2
#define SST25_CS_LAT        _LATC2
#define ERASE_SECTOR_SIZE   65536
#define SPI_FLASH_CONFIG    { 2, 3, 6, 0, 1, 1, SPI_MODE_8BITS}
```

Listing 5.15 – *HardwareProfile.h* **Serial Flash support section**

The first define is responsible for the selection of the M25P80 specific set of commands – one of the key differences between the various serial flash families.

The second define is used to index the specific SPI peripherals used.

The following two lines specify the I/O pin responsible for the control of the *Chip Select* line. Note the `SST25` prefix being used here as well.

The `ERASE_SECTOR_SIZE` constant is defined here as it is one of the key characteristics of this particular serial flash memory model.

Finally the `SPI_FLASH_CONFIG` symbol helps us configure the *drv-spi.c* module setting the correct prescaler value, clock polarity and active edge for the microcontroller SPI port.

The new *main.c* module will start with the usual list of includes where we will add the *m25p80.h* for the first time.

A global variable, an array of char named `data` is defined here as a buffering mechanism.

The `main()` function contains the usual initialization calls required to set up the I/O port and to initialize the LCD terminal emulation and grid based touchscreen input.

A splash screen is also provided as in the previous examples, before the actual main loop begins.

```c
/*
 * Project: Storage_flash
 *
 * File:    main.c
 *
 * Requires: MLA 1306
 */
#include "PICconfig.h"

#include "TimeDelay.h"
#include "LCDTerminal.h"
#include "TouchGrid.h"
#include "MDD File System/FSIO.h"
#include "menu.h"
#include "M25P80.h"

char data[ 4096];           // ensure ERASE_SECTOR_SIZE is divisible by sizeof(data)!

int main( void )
{
    FSFILE *fp;
    unsigned length;
    char *p, filename[32];
    DWORD address, i, scount;

    // 1. initializations
    uMBInit();                  // init pins and ports
    LCDInit();                  // inits terminal emulation
    DisplayBacklightOn();
    TouchGridInit( 3, 3);       // defines a 3x3 grid

    // 2. splash screen
    LCDClear();
    LCDCenterString( -1, "Serial Flash demo");
    LCDCenterString( +1, "tap to start");
    TouchGrid();
```

Listing 5.16 – Storage_flash, main.c, first part

The main loop, in Listing 5.17, is where things start getting interesting.

`SelectFile()` is called to allow the user to pick the file to be used for the demonstration among those with the extension "*.TXT". The chosen file is opened for reading using `FSfopen()` and, if successful, the actual copying process takes place.

Before writing to the serial flash memory, it is necessary to operate first an erase of the entire sector where the data will be written to.

There are many ways to keep track of when we need to perform a flash sector erase and where we are along the process of writing, but for simplicity we will use two variables of adequate size: *address* and *scount*. They are defined as DWORDS (32-bit unsigned integers) to make sure that they can correctly track quantities that could be in excess of the typical 16-bit integer range of the "`int`" type used by the MPLAB XC compiler.

```c
// 3. Main Loop
while( 1 )
{
    LCDClear();

    // 4. try to open a file
    SelectFile( filename, "*.TXT");
    LCDClear();

    // 5. open file
    fp = FSfopen( filename, "r");
    if ( fp == NULL)
    {
        LCDCenterString( 0, "Cannot open file");
    }
    else // copy
    {
        address = 0;        // absolute address
        scount = 0;         // sector relative count
        LCDClear();

        // 6. copy file in flash
        while ( !FSfeof( fp))
        {
            // 7. erase sector at the beginning
            if ( scount == 0)
                SST25SectorErase( address);
            LCDPutString( "\n***Erasing Sector***");

            // 8. copy (up to) one sector of data at a time
            length = FSfread( &data, 1, sizeof( data), fp);
            LCDPutString( "\n***Writing Data***");

            // 9. write to device
            SST25WriteArray( address, data, length);
            address += length;

            // 10. roll over the counter within a sector
            scount  += length;
            if ( scount > ERASE_SECTOR_SIZE)
                scount = 0;
        }
        FSfclose( fp);

        // 11. read back and display on terminal
        LCDPutString( "\n ***Enf of File***");
        LCDPutString( "\n Tap to read back");
        TouchGrid();

        for ( i=0; i<address; i++)
            LCDPutChar( SST25ReadByte( i));

        LCDPutString( "\n Enf of File");
    } // copy

    // 12. prompt to continue
    TouchGrid();

    } // main loop
} // main
```

Listing 5.17 – Storage Flash, *main.c*, main loop

The `address` variable is used to keep track of the writing process and keeps incrementing indefinitely at each block of data written.

The `scount` variable is periodically reset to keep track of the erase sector blocks and each time this triggers a new sector erase command.

For the correct operation of this simplified mechanism it is necessary that the buffer (data) size be a value that can fit an exact number of times in the erase-sector size. In other words the erase sector size must be divisible by the buffer size!

The *data* buffer can now be repeatedly filled with new data read in bulk (using `FSfread()`) from the SD card file and from the buffer it can be written to the Serial Flash using the `SST25WriteArray()` command.

When the end of file is reached (`FSfeof()` returns true) the copy process is complete, the input file is closed and a complete read back is performed form the Serial Flash memory one byte at a time (for demonstration purposes) using the `SST25ReadByte()` function.

After offering a prompt and waiting for the user touch, the whole operation can be repeated all over again, possibly selecting a new and different file from the SD card.

Build the project and run it on the Mikromedia board to verify that both the read and write operations are performed correctly on the Serial Flash memory.

Note that for maximum simplicity and portability across various Serial Flash models, the `SST25WriteArray()` and `SST25ReadArray()` functions are not taking advantage of any of the specialized commands available in the M25P80 device. As a consequence the performance that can be expected is far from optimal.

NOTE

> Numerous inline comments in the *m25p80.c* source file do make reference to architectural details of the SST25 devices, and should be ignored to avoid confusion. One such detail is the erase-sector size that on the SST25 series devices is only 4096 bytes large. Should this value be used incorrectly in an algorithm applied to the M25p80 device instead (such as the copy loop in our demo project), it would result in a repeated erase of the same (in truth larger) sector with consequent loss of all but the last 4096 bytes segment copied into it.

Tips & Tricks

Using Serial Flash to store TouchScreen Calibration Data

Now that we have learned to include the Serial Flash memory in our applications, we can put it to use in another context, the touch screen support module.

In the previous chapter we have been initializing the touch screen module without using any non-volatile memory support, hence we had no way to store the screen calibration data and the procedure had to be repeated after each restart or we had to force the touchscreen module to use pre-set values hardcoded in the application.

We have now the option to rectify the situation by *spending* one of the sectors of the serial flash (the last one is a common choice) and allowing the touch screen module to use the it to store the calibration data record.

This can be considered quite a high price to pay as we will be effectively storing just a few dozens of bytes of data in what is a rather large sector of flash memory. Eventually the decision is ours and the trade off is as usual one between "cost" and "convenience".

Practically speaking, adding non volatile storage of the touch screen calibration data requires that we modify only the `TouchInit()` function call (in *TouchGrid.c* for example) by providing pointers to appropriate *write, read* and *erase* functions as the first three parameters (previously set to `NULL`) :

```
// initialize the touch state machine
TouchInit( &SST25WriteWord, &SST25ReadWord, &SST25SectorErase, NULL);
```

We must also include the proper I/O initializations for the Serial Flash device *Chip Select* signal:

```
// initialize the Serial Flash
SST25_CS_LAT = 1;
SST25_CS_TRIS = 0;
SST25Init( &si );
```

If we include this code directly into the `uMBInit()` function (in the *uMedia.c* module), we can also take the opportunity to include the `FlashInit()` function call so to make access to the large serial flash immediately available :

```
DRV_SPI_INIT_DATA spi_config = SPI_FLASH_CONFIG;
FlashInit( &spi_config );
```

Summary

In this chapter we have explored storage options available on the Mikromedia board and how the MLA libraries support them.

The micro SD card slot, thanks to the MDD File System library is by far the most powerful tool available to us to access Gigabytes of data. The Image Decoding library allows us to overcome the on chip flash memory space limitations and access almost unlimited image files to build our user interfaces.

The serial flash device, on the other hand, brings a complementary set of options. While smaller in size (1MB) and limited to only 16 large sectors of data individually erasable, it can offer higher reliability and the advantage of a permanent connection. When used in conjunction with the touch screen module, it can also provide us with convenient non volatile storage for the screen calibration.

Suggested Reading

- Di Jasio - **"Programming 16-bit microcontrollers in C" - Newness Elsevier**

    67. Chapter 13 – Mass Storage, learn about low level SD/MMC card access

    68. Chapter 14 – File I/O, learn about the FAT16 file format

    69. Chapter 7 – Synchronous Communication, learn about the SPI interface and access to EEPROM access devices

Exercises

1. Modify the Text example to produce the *hex dump* of a generic binary file.
2. Modify the Slides example to turn it into a slide show application.
   Also add visual aids (arrow symbols) to suggest the user where to touch.
3. Create a function to save a screen snapshot to a file on the micro SD card.

Solution to Exercise 3

```
void ScreenCapture( char *filename)
{
    FSFILE *fp;
    GFX_COLOR Row[ 320];
    int i, j;

    // open file
    fp = FSfopen( filename, FS_WRITE);
    if ( fp != NULL)
    {
        // dump contents of the screen
        for(j=0; j<=GetMaxY(); j++)
        {
            // row by row
            for( i=0; i<=GetMaxX(); i++)
            {
                Row[ i] = GetPixel( i, j);
            }

            // write buffer to file
            FSfwrite( Row, sizeof(Row), 1, fp);
        }
        // close file
        FSfclose( fp);
    }
}
```

Listing 5.18 - ScreenCapture() function

This function can be conveniently included in the *uMedia.c* support module so to be available to all projects. It can be invoked at any time when a quick screen snapshot is required as long as the MDD File System support files are linked in and the library has been initialized (`FSInit()`).

Note that the output file contains the array with the pixel color values encoded as binary 16-bit values and will not be recognized as a standard image file by GIMP or other graphic editors.

Once the micro SD card is inserted in a PC reader a small Python script (see Listing 5.19) can be used to reproduce the image captured onto the computer screen. From there it will be easy to cut and paste it using any of the native operating system tools (Windows Printscreen, OS X ScreenCapture ...)

```
#
# ScreenCapture visualizer
#
import sys
import pygame

# init graphics environment
pygame.init()

# define a screen surface
size = width, height = 320, 240
screen = pygame.display.set_mode( size)

# main loop
while 1:
    for event in pygame.event.get():
        if event.type == pygame.QUIT:
            sys.exit()

    # paint image
    screen.fill( (0, 0, 0))

    if len(sys.argv) == 0:
     filename = "SCREEN.SCR"
    else:
        filename = sys.argv[1]
    try:
        f = open( filename, "rb")
        for j in range( 240):
            for i in range( 320):
                bytel = ord( f.read(1))
                byteh = ord( f.read(1))
                b = ((bytel & 0x1f) << 3) + 7
                r = byteh & 0xF8 + 7
                g = ((bytel & 0xe0) >> 3) + ((byteh & 0x7) << 5) + 3

                pygame.draw.rect( screen, ( r, g, b), ( i, j, 2, 2), 0)
    finally:
        f.close()

    pygame.display.flip()
```

Listing 5.19 - ScreenCapture.py a Python Script to reproduce captured images

Online Resources

- **Python for Windows** – http://www.python.org/getit/windows/

    Note that Python comes pre-installed on OS X and most Linux distributions

- **Pygame** – http://www.pygame.org

    A Python library that provides basic cross platform graphics support

Chapter 6

Sound

In this chapter we will take a look at the reproduction of sounds and music using the Mikromedia board audio interface. As we will see this is composed of a powerful audio codec chip that is capable of handling directly a wide variety of audio file formats. First we will test the interface taking advantage of basic debugging features that are built in the decoder chip. Later, we will explore the possibility of playing back sound clips (audio resources) stored in the chip flash memory and eventually combine the capabilities of the MDD File System library to perform playback of audio files directly from an SD card.
In the end we will obtain a small support module that will integrate with the rest of the MLA libraries adding the sound dimension to our applications.

The VS1053 codec

The PIC24 Mikromedia board features one of the most popular and easy to use codec chips currently available: the VS1053 from VLSI technologies. This device is part of a long lineage of codec chips that have been dominating the consumer and embedded control market for the last decade. At every step along this brief history they have been refined and new features have been added but, since the VS1002 model, the microcontroller interface has remained for the most part unchanged allowing for a great deal of software portability across models.
The VS1053 in particular offers:

- Dual Synchronous Serial interfaces (for data and command) with the possibility to use and share a single SPI interface and a minimal number of I/Os.

- Amplified stereo outputs that can drive directly headphones or a pair of small loudspeakers

- A powerful DSP core that is capable of interpreting directly the contents of MP3, WMA, MIDI (with 128 instruments), Ogg Vorbis, AAC file formats, but also WAV (RIFF) which includes the most basic PCM and ADPCM modes and a list of 20 additional variants.

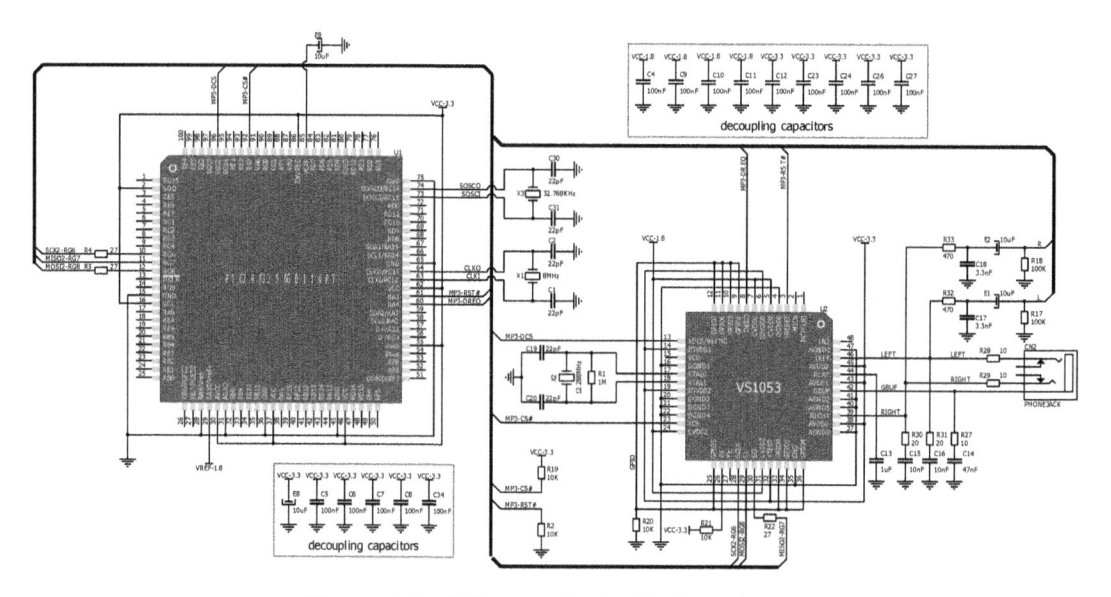

Figure 6.1 – Mikromedia Audio Interface

Figure 6.1 illustrates the connections provided on the PIC24 Mikromedia board which include the SPI2 interface (SDI, SDO and SCK) and four additional I/O pins. Here is a detailed description of the four additional I/O pins and their connections to the microcontroller specific I/O:

- *MP3-Reset*, (PORTA pin 5, output) to force a hard reset of the codec chip
- *MP3-DCS*, (PORTA pin 3, output) to select the SPI Data streaming interface
- *MP3-CS*, (PORTA pin 2, output) to select the SPI Command interface
- *MP3-DREQ*, (PORTA pin 4, input) to detect when the codec is ready to receive more data

NOTE

The VS1053 pin names reflect the first and foremost use of the codec chip in the designers' mind which was that of reproducing MP3 files, although as we know, many more file formats are supported.

There are in fact a few more interesting features offered by the VS1053 chip that are not used by the Mikromedia board but could be of interest in future applications, among which:

- An analog input, meant for a microphone input that can be used to encode on the fly music into a small set of file format options

- I2S digital output interface, that can be used to connect (chain) other codecs or DSPs to further process the output signal (or to stream it wirelessly...)
- UART interface for "real time" MIDI execution

Serial Command Interface

The VS1053 codec requires two separate interfaces, one for data and one for commands, but smartly multiplexes them over the same synchronous serial port (SPI):
- When the *XCS* line is active (low), the SPI port is connected to the *Command interface*.
- When the *XDCS* line is active (low) the SPI port is connected to the *Data interface*.

The serial command interface (often referred to as *SCI* in the VS1053 literature) is designed around two simple commands (READ and WRITE) that can access a small number of 16-bit wide control registers.

Each command is therefore composed of a sequence of four bytes that need to be sent to the codec with the traditional SPI convention – most significant bit first.
- The first byte encodes the command itself:

Command	
Name	Opcode
WRITE	0x02
READ	0x03

Table 6.1 – Serial Command Interface commands

- The second byte is the address of the desired 16-bit control register.
- The third and fourth bytes contain the data being transferred (MSB first).

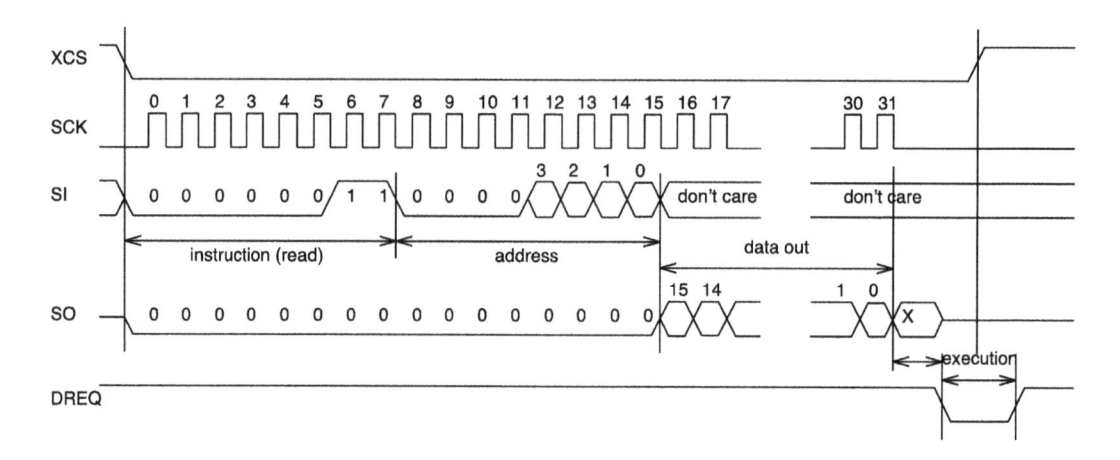

Figure 6.2 – Word Read Command

As illustrated in Figure 6.2, the sequence must be preceded by lowering the *XCS* line (*MP3_CS* on the Mikromedia board schematic), rising it again once the last bit of the command sequence has been shifted out.

During use of the command interface the *DREQ* line indicates whether the codec is busy (low) executing a command or is ready to receive a new one (high).

There are then 16 registers that can be addressed, written to and read from (two of them are read only actually) among which:

- *MODE* register (0x00), which contains 16 individual control bits each responsible for enabling or disabling a specific feature of the device
- *STATUS* register (0x01), which provides an important window to the inner working of the device and its current state
- *VOL* register (0x0B), which controls the output volume

It is beyond the scope of this chapter to examine each and every command and register in detail, for which I refer you to the original device user manual, but we will develop in the following a small set of functions that will allow us to exercise the most common features of the device.

VS1053.h

In fact we can start right away creating a new source (header) file: **VS1053.h** to capture the basic commands and register names. See Listing 6.1.

Save this file in the shared folder (*uMB*) so that we will be able to use it in all future projects.

```c
/*
 * File:    VS1053.h
 *
 */
#ifndef _VS1053
#define _VS1053

#define USE_AND_OR

#include <ports.h>
#include <TimeDelay.h>
#include "HardwareProfile.h"

// basic SCI commands
#define MP3_CMD_WRITE       0x02
#define MP3_CMD_READ        0x03

// basic registers list
#define MP3_REG_MODE        0x00    // mode options
#define MP3_REG_STATUS      0x01    // main status reg
#define MP3_REG_BASS        0x02    // enable bass & treble enhancements
#define MP3_REG_CLOCKF      0x03    // clock + freq. multiplier
#define MP3_REG_DECODE_TIME 0x04    // in seconds
#define MP3_REG_AUDATA      0x05    // misc. audio data
#define MP3_REG_WRAM        0x06    // RAM write/read
#define MP3_REG_WRAMADDR    0x07    // base address for RAM write/read
#define MP3_REG_HDAT0       0x08    // stream header data 0
#define MP3_REG_HDAT1       0x09    // stream header data 1
#define MP3_REG_AIADDR      0x0A    // start address of application
#define MP3_REG_VOL         0x0B    // volume control

// MODE register options
#define MP3_MODE_DIFF       0x01    // differential (left channel inverted)
#define MP3_MODE_LAYER12    0x02    // allow MPEG layers 1 & 2
#define MP3_MODE_RESET      0x04    // soft reset
#define MP3_MODE_OUTOFWAV   0x08    // jump out of WAV
#define MP3_MODE_TEST       0x20    // allow test sequences
#define MP3_MODE_STREAM     0x40    // stream mode
#define MP3_MODE_DACT       0x100   // DCLK active edge (rising, default: falling)
#define MP3_MODE_SDIORD     0x200   // SDI bit order (default: MSb first)
#define MP3_MODE_SDISHARE   0x400   // share SPI chip select
#define MP3_MODE_SDINEW     0x800   // new VS1002 native mode

#endif
```

Listing 6.1 – VS1053.h header file

Hardware Profile VS1053 Decoder Section

As with previous board features, we can add a specific section to the *HardwareProfile.h* header file to help us abstract the specific interface details.

```
/************************************************************************
* IOS FOR THE VS1053 decoder
*************************************************************************/
// SPI2 is shared with uSD card in "VS1002 native mode"

#define MP3_RST_Config()        _TRISA5 = 0 // o reset decoder
#define MP3_RST_Enable()        _LATA5 = 0
#define MP3_RST_Disable()       _LATA5 = 1

#define MP3_DREQ                _RA4        // i request for data

#define MP3_DCS_Config()        _TRISA3 = 0 // o data select
#define MP3_DCS_Enable()        _LATA3 = 0
#define MP3_DCS_Disable()       _LATA3 = 1

#define MP3_CS_Config()         _TRISA2 = 0 // o command select
#define MP3_CS_Enable()         _LATA2 = 0
#define MP3_CS_Disable()        _LATA2 = 1

#define MP3_SPICON1             SPI2CON1
#define MP3_SPISTAT             SPI2STAT
#define MP3_SPIBUF              SPI2BUF
#define MP3_SPIRBF              SPI2STATbits.SPIRBF
#define MP3_SPIENABLE           SPI2STATbits.SPIEN
```

Listing 6.2 – VS1053 section added to HardwareProfile.h

Serial Data Interface and Protocol

By selecting the *XDCS* line (active low), we get access to the main data FIFO buffer of the codec which is 2048 bytes large. During the playback it is our job to keep this buffer from ever becoming empty, to avoid gaps in the reproduction, but we must also avoid overflowing it. So it is important that we periodically check the *DREQ* line which will tell us if the buffer is full or if it can accept more data. When *DREQ* is high, we can be sure that there is room to send at least 32 bytes of new data. Note that we don't need to check the state of this line after each byte sent. We can proceed sending an entire packet of up to 32 bytes without further ado.

The data to send will be the MP3 (or otherwise formatted) compressed contents of our audio *stream* be it a sound sample, a song or a MIDI command. In any case the entire contents of the *stream* will be sent, including various headers and optional tags as present on the media source. In fact it will be those headers that will help the VS1053 determine what kind of data is arriving and to choose the appropriate decoding algorithm and playback sample rates.

Preparation

As in the previous chapter, we will create a new project and immediately populate it with the key components of the Graphics library and board support package. Here are the simple steps:

1. Use the **New Project Wizard** to create a new project inside the working directory (*Mikromedia*), let's call it: "**6-Audio_Test**"

2. Add a new logical folder to the Source Files folder, called **MLA**

3. Add the following items to the folder:

    - **TimeDelay.c**, a few basic timing (blocking) functions used in the driver

4. Add a new logical folder to the Source Files folder, called **uMB**

5. Add the following item to the folder:

    - **uMedia.c**, to provide the Mikromedia board (PPS) initialization functions

6. Configure the compiler **C include dirs** path to contain:

    - **. (dot)**, the current project directory for MLA to reach our configuration files

    - **../Microchip/Include**, for our source files to reach inside the MLA

    - **../uMB**, for project source files and the MLA to access the Hardware Profile and other resources shared among this and future projects in this book and specific to the Mikromedia board.

7. Create a new *main.c* file using the **New File wizard** and the embedded template or your own customized version

Initialization and Serial Communication

Listing 6.3 shows the first two functions that we need to send data through the SPI port and to select a particular command register.

```
/*
 * Project:    6-Audio_Test
 *
 * File:       main.c
 */
#include "PICconfig.h"

#include "TimeDelay.h"
#include "uMedia.h"
#include "VS1053.h"

WORD writeMP3( BYTE b)
{
    MP3_SPIBUF=b;                   // move byte into buffer
    while(!MP3_SPIRBF);             // wait for transmission completion
    return MP3_SPIBUF;              // return buffer content
}

void writeMP3Register( BYTE reg, WORD w)
{
    MP3_DCS_Disable();              // disable data bus
    MP3_CS_Enable();                // enable command bus
    writeMP3( MP3_CMD_WRITE);
    writeMP3( reg);
    writeMP3( w>>8);                // MSB first
    writeMP3( w & 0xff);
    MP3_CS_Disable();
    while( !MP3_DREQ);
}
```

Listing 6.3 – Audio_Test, main.c, Writing to the SPI interface and Control Registers

Notice how we used the macros provided in the new section of the Hardware Profile to make the code independent from the specific I/O pins and SPI port selected.

Using `writeMP3()` we can send data both to the serial command and the serial data interface. Enabling the *MP3_CS* or the *MP3_DCS* we will be able to steer the information to the desired interface.

The function `writeMP3Register()`, instead is specific to the Serial Command Interface and packs the four bytes of information as required by the WRITE command protocol.

We can also proceed to code the main initialization function: `MP3Init()` that will perform the following simple steps:

1. Configure the I/Os assigned to control the codec
2. Configure the SPI port in a compatible mode
3. Hard reset of the device, holding the *MP3_RST* line down for at least 10ms
4. Release the reset line and wait for additional 10ms to allow the codec to initialize properly

Sound - 163

5. Set the *MODE* register with a default configuration (known as *SDI_NEW*) and additional flags passed (optionally) in the *mode* parameter
6. Wait for the codec to complete its initializations by polling the *DREQ* line

```
void MP3Init( WORD mode)
{
    volatile int temp;

    /// PPS must be initialized
    MP3_RST_Enable();       // force reset
    MP3_CS_Disable();
    MP3_DCS_Disable();
    MP3_RST_Config();       // make Reset output
    DelayMs( 1);
    MP3_CS_Config();        // make xCS output
    MP3_DCS_Config();       // make xDCS output

    // init SPI slow
    MP3_SPIENABLE = 0;      // spi off
    MP3_SPICON1 = 0x007C;   // cke=0 and ckp=1, master 8-bit, pre 1:64
    MP3_SPIENABLE = 1;      // on

    DelayMs( 10);           // give time to reset
    MP3_RST_Disable();      // release reset

    // verify that codec is busy resetting
    while ( MP3_DREQ);

    //verify that codec is ready to receive first command
    while( !MP3_DREQ);

    // set native + mode
    writeMP3Register( MP3_REG_MODE, MP3_MODE_SDINEW | mode);
    // double the codec clock frequency
    writeMP3Register( MP3_REG_CLOCKF, 0x2000);

    // change SPI to max speed
    if ( mode == 0)
    {
        MP3_SPIENABLE = 1;    // spi off
        MP3_SPICON1 = 0x007E;
        MP3_SPIENABLE = 1;    // on
    }

} // MP3Init

WORD readMP3Register( BYTE reg)
{
    WORD w;

    MP3_DCS_Disable();              // disable data bus
    MP3_CS_Enable();                // enable command bus
    writeMP3( MP3_CMD_READ);
    writeMP3( reg);
    w = writeMP3( 0xff) ;           // write dummy, get MSB first
    w = (w<<8) | writeMP3( 0xff);   // write dummy, get LSB
    MP3_CS_Disable();
    while( !MP3_DREQ);
    return w;
}
```

Listing 6.4 – Audio_Tes, maint.c, Initializing the Codec and Reading a Control Registers

Note that the `MP3Init()` function (in Listing 6.4) assumes that the SPI port of choice has already been connected to the correct I/Os, which means we must initialize the Peripheral Pin Select module of the microcontroller using `uMBInit()` before calling it.

Initially we set the SPI port clock to a safe slow mode (prescaler 1:64). After the clock of the codec has been correctly configured (enabling a PLL circuit to double the clock frequency) we can rise the SPI port speed to its maximum (prescaler 1:1).

Finally the `readMP3Register()` function allows us to read back the contents of a control registers.

This could be in fact our first test (reading back the contents of the *MODE* register) after initialization to verify that the codec is indeed set in the mode we specified.

Sine Test

But a far more telling (and rewarding) test can be performed using one of the diagnostic functions of the VS1053, the so called: *sine test*.

The *sine* test is built into the VS1053 and can be activated sending a particular *data* string directly to the FIFO, through the Serial Data Interface.

The string is composed of the following 8 bytes: 'S', 0xEF, 'n', n, 0, 0, 0, 0
(encoded in ASCII as `0x53, 0xEF, 0x6E, n, 0x00, 0x00, 0x00, 0x00`)

Where *n* is a parameter that defines the particular sample rate and resulting output sinusoid frequency to be used. This parameter is actually split in two bit fields:

- The first one (composed of bit 7,6 and 5) forms an index that selects a sampling frequency among the values in Table 6.2:

Index	Sampling Frequency Fx
0	44,100
1	48,000
2	32,000
3	22,050
4	24,000
5	16,000
6	11,025
7	12,000

Table 6.2 - Sampling Frequency Selection

- The remaining 5 bits (bit 4,3,2,1,0) form a coefficient *S* that is entered in the following equation to determine the output sinusoid frequency:

$$F = Fx * S / 128$$

To generate a steady 1kHz tone, for example, we can enter the value $n = 0x44$.
This splits in the two bitfields: 2 and 4 ,therefore selecting:

- Fx = 32kH sampling frequency

- S = 4 coefficient resulting in F = 32,000Hz * 4 / 128 = 1,000Hz

```
void testMP3Sine( BYTE n)
{
    // Send a Sine Test Header to Data port
    MP3_DCS_Enable();         // enable data bus
    writeMP3( 0x53);          // special Sine Test Sequence
    writeMP3( 0xef);
    writeMP3( 0x6e);
    writeMP3( n);             // n, Fsin = Fsamp[n>>5] * (n & 0x1f) / 128
    writeMP3( 0x00);          // where Fsamp[x] = {44100,48000,32000,22050,
    writeMP3( 0x00);          //                   24000,16000,11025,12000}
    writeMP3( 0x00);          // for example n = 0x44 -> 1KHz
    writeMP3( 0x00);
    MP3_DCS_Disable();

    DelayMs( 500);            // continue sinusoid for 500ms

    // Stop the sine test
    MP3_DCS_Enable();         // enable data bus
    writeMP3( 0x45);          // special Sine Test termination sequence
    writeMP3( 0x78);
    writeMP3( 0x69);
    writeMP3( 0x74);
    writeMP3( 0x00);
    writeMP3( 0x00);
    writeMP3( 0x00);
    writeMP3( 0x00);
    MP3_DCS_Disable();

    DelayMs( 500);
} // Sine Test
```

Listing 6.5 – Sine Test function

The `testMP3Sine()` function in Listing 6.5 allows us to enter any desired value for the parameter 'n' in the sequence and then waits for half a second before terminating the test with the string: 'E', 'x', 'i', 't', 0, 0, 0, 0
(encoded in ASCII as `0x45, 0x78, 0x69, 0x74, 0x00, 0x00, 0x00, 0x00`)

At this point you might be wondering how likely it is that any such short sequence of characters could appear inside a given audio file. The actual probability is pretty low, given the continuous nature of an audio signal and the somewhat abrupt discontinuity in the two strings produced by the four zeros, but it is a finite number!
To prevent unintended activations of this and other diagnostic functions, the recognition of such test sequences can be enabled only by entering a particular diagnostic mode by means of setting a dedicated bit (`MP3_MODE_TEST`) in the *MODE* control register of the codec.

This is precisely the reason why we allowed for a parameter to be passed to the `MP3Init()` function.

```
void main( void )
{
    // initializations
    uMBInit();

    MP3Init( MP3_MODE_TEST);        // TEST MODES enabled

    while( 1 )
    {
        testMP3Sine( 0x44);         // 32ksmps, 1KHz out, 500ms pulse
    } // main loop
}
```

Listing 6.6 – Audio Test, main() function

In Listing 6.6, we can see how the `main()` function makes use of the `uMBInit()` function (found in the support module *uMedia.c*) to configure the Mikromedia board I/Os (establishing the correct Peripheral Pin Select connections between the SPI2 port and the standard I/O pins) before invoking the `MP3Init()` function with the *TEST* mode bit set. In the main loop we call `testMP3Sine()` to produce a pulsed 1kHz tone with a period of 1Hz and 50% duty cycle.

You can verify the presence of the tone plugging a headset in the 3.5mm stereo jack of the Mikromedia board, but unless you have a 'perfect pitch' you will need a frequency meter or, even better, an oscilloscope to prove the exact frequency output.

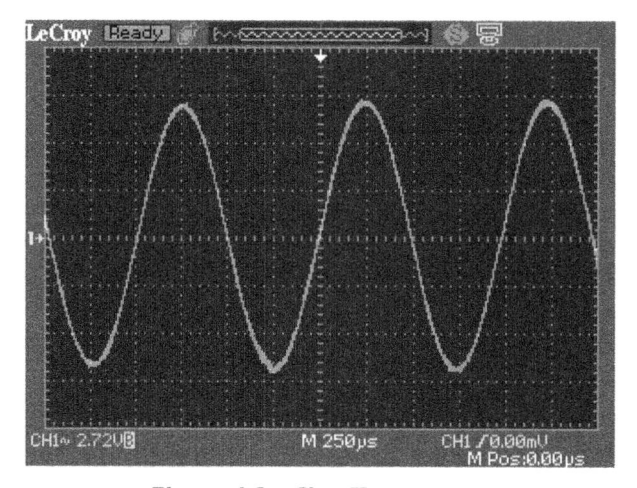

Figure 6.3 – Sine Test output

Hello Again!

With just a few more lines of code we will soon be ready to produce a much more impressive audio demonstration by playing a real MP3 encoded audio sample.
All we need is an efficient mechanism to help us feed a constant flow of data to the codec. The Serial Data Interface protocol is based on the use of the *DREQ* flag to indicate when the codec FIFO can accept a new batch of samples. The flag, in particular, is raised when the FIFO has room for at least 32 bytes. So it makes sense to develop a feeding function that operates in packets of 32 bytes at a time.

```
void simpleFeedMP3( BYTE *pdata, size_t length)
{
    int i;
    while ( MP3_DREQ && ( length > 0))    // FIFO has room and data is available
    {
        MP3_DCS_Enable();
        for( i = 0; i < 32; i++ )          // send a packet of up to 32-bytes
        {
            writeMP3( *pdata++);           // send data
            length--;                      // decrement counter

            if( length == 0)               // exit when source exhausted
                break;
        }
        MP3_DCS_Disable();
    }
} // simple feed MP3
```

Listing 6.7 – simpleFeedMP3() function

Each time we call `simpleFeedMP3()` in Listing 6.7, it performs a test to see if the FIFO is ready to accept data. If so it sends a block of 32-bytes and repeats until the FIFO cannot take more 32-byte packets or our source has been consumed entirely.

NOTE

> If the FIFO is initially empty, this function will continue to feed data to the codec until full (all 2048 bytes) or the source is exhausted (if shorter than 2,048 samples). With the SPI port set for maximum speed (16MHz) this means that the function might not return for as long as 1ms after the first call.

A more useful version of this function can be derived (Listing 6.8) if we pass the two parameters, the p*data* pointer and data *length,* via pointers. This *double indirection* makes the code a bit less readable but makes the resulting `feedMP3()` function more flexible as we can now advance the pointer and decrement the counter between calls:

```
void feedMP3( BYTE **ppdata, size_t *plength)
{
    int i;
    while ( MP3_DREQ && (*plength > 0))    // while FIFO has room and data available
    {
        MP3_DCS_Enable();
        for( i = 0; i < 32; i++ )          // send a packet of up to 32-bytes
        {
            writeMP3( *(*ppdata)++);        // send data
            (*plength)--;                   // decrement counter
            if( (*plength) == 0)            // exit when source exhausted
                break;
        }
        MP3_DCS_Disable();
    }
} // feed MP3
```

Listing 6.8 – feedMP3() function

Let's create a new demonstration project by copying the contents of the previous *Audio_Test* project in a new project (and folder) that we will call: ***Audio_Hello.***

Creating VS1053.c

The contents of the *main.c* file can be re-used now to create the ***VS1053.c*** source file that we will save in the shared folder (*uMB*). We will discard the `main()` function and possibly the `testMP3Sine()` but we will add the `feedMP3()` function and two new short ones:

```
void flushMP3( void)
{
    int i;
    MP3_DCS_Enable();
    for( i = 0; i < 2048; i++ )
    {
        while( !MP3_DREQ );
        writeMP3( 0x00 );
    }
    MP3_DCS_Disable();
} // flush MP3

void setMP3Volume( WORD left, WORD right)
{
    writeMP3Register( MP3_REG_VOL, (left<<8) + right);
} // set MP3 Volume
```

Listing 6.9 – VS1053.c, flushMp3() and setMP3Volume()

The `flushMP3()` function is required, according to the codec documentation, to ensure that the DSP engine terminates correctly the decoding of the last packet of data in a given audio sample. It basically fills the buffer with zeros ensuring that every last bit of information in it contained has been played back.

The `setMP3Volume()` function, sends simultaneously two new volume values to the left and right channel control registers. The two parameters, *left* and *right,* are actually 8-bit values allowing for 256 levels of volume control, but get combined in a single 16-bit word sent to the *VOL* register (0x0B).

We can now add the complete set of prototypes to the *VS1053.h* header file:

```
// prototypes
unsigned  writeMP3( BYTE b);
void writeMP3Register(BYTE reg, WORD w);
unsigned readMP3Register( BYTE reg);

void MP3Init( WORD mode);
void setMP3Volume( WORD left, WORD right);
void testMP3Sine( BYTE n);
void feedMP3( BYTE **pdata, size_t *plength);
```

Listing 6.10 – VS1053.h prototypes

Now that we have everything we need for a proper playback, we are going to need an audio sample encoded in MP3. The simplest way to supply it is to include it as an *audio resource*, that is, include it as an array of bytes properly initialized in flash memory. But, typing manually, if only a few thousand hexadecimal values, can be a prohibitively error prone and tedious work. I highly recommend instead that we use **Audacity®**, a powerful open-source and cross-platform audio editing tool, to perform the capture, isolate the audio segment desired and, thanks to an extension module called **LAME**, to perform the MP3 encoding.

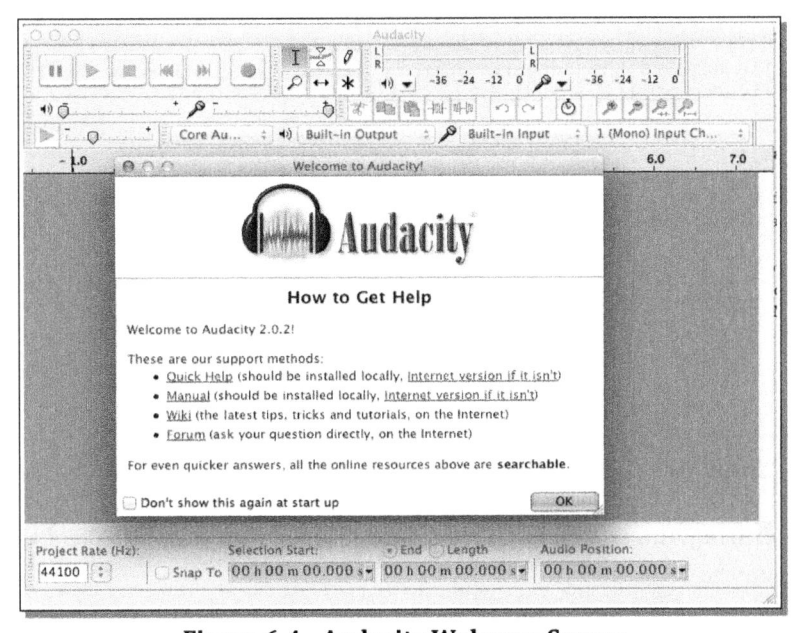

Figure 6.4 - Audacity Welcome Screen

Eventually, we can use a simple Python script to convert the MP3 file obtained into a C array declaration and initialization, a resource file ready to be included in our project.

```python
#
# Convert MP3 files to C arrays
#
# usage: python MP3toC.py name[.mp3] > name.h
#    or: ./MP3toC.py name[.mp3] > name.h
#
import sys

# identify source file
argc = len( sys.argv)

if argc != 2:
    print "usage ./MP3toC.py name[.mp3] > name.h"
    sys.exit( 1)

source = sys.argv[1]
source_list = source.split('.')

# check if an extension was given, otherwise append .mp3
if len( source_list) == 1:
    source += '.mp3'

try:
    fsource = open( source, "rb")
except IOError:
    sys.stderr.write( 'File %s not found!\n' % source)
    sys.exit( 1)

# produce include file heading
print """
/*
 * audio resource %s
 */

// initialized data
const BYTE """ % source,

# use the file name as the resource name
print '_' + source.split(".")[0] + "[] = {"

try:
    while( 1):
        b = fsource.read(16)
        for c in b:
            print "%#0.2x," % ord(c),
        print
        if len( b) < 16:
            print "};"
            break

finally:
    fsource.close()

print
```

Listing 6.11 – Python script to convert an MP3 file into C array declaration

You can try capturing and encoding your own message or you can download the example file *Hello.h* from the book web site.

```
/*
 * audio resource hello.mp3
 *
 */

// initialized data

const BYTE _Hello[] = {
0xff, 0xfb, 0x50, 0xc4, 0x00, 0x00, 0x00, 0x00, 0x00, 0x00, 0x00, 0x00, 0x00, 0x00, 0x00,
0x00, 0x00, 0x00, 0x00, 0x00, 0x49, 0x6e, 0x66, 0x6f, 0x00, 0x00, 0x00, 0x0f, 0x00, 0x00, 0x00,
0x25, 0x00, 0x00, 0x1f, 0x04, 0x00, 0x06, 0x06, 0x0d, 0x0d, 0x0d, 0x14, 0x14, 0x14, 0x1b, 0x1b,
...
};
```

Listing 6.12 – Hello.h, MP3 Encoded Audio Sample

Listing 6.12 provides a snippet of the resource file produced by the *MP3toC.py* script. In it we can see the definition of the `_hello[]` array and the first few lines of its long initialization { ... } vector containing the MP3 encoded audio stream.

The new *main.c* file (see Listing 6.13), beside the usual includes, adds the newly completed *VS1053.h*, and the resource file *Hello.h*

The `main()` function performs the usual initialization of the microcontroller I/Os followed by the initialization of the codec. There is no need to enable the test mode this time, but we need to set the audio volume to a mid-low level to avoid a "surprise" later when testing the application.

In the main loop, we initialize the pointer to the audio sample array (1.), we take its length, and we invoke the `feedMP3()` function repeatedly (2.) until it has been all sent to the FIFO.

The `flushMP3()` function will ensure that the entire contents of the FIFO are played back. Eventually, after a short pause (3.) we repeat the test.

```c
/*
 * Project: Audio_Hello
 *
 * File:    main.c
 */
#include "PICconfig.h"

#include "TimeDelay.h"
#include "uMedia.h"
#include "vs1053.h"

#include "Hello.h"                  // include the audio resource

int main( void )
{
    size_t length;
    BYTE *p;

    // initializations
    uMBInit();                      // init PPS

    // init the MP3 player
    MP3Init( 0);                    // init MP3 decoder
    setMP3Volume( 30, 30);          // set volume

    // Main Loop
    while( 1 )
    {
        // 1. init pointer to flash audio resource
        p = (void*) _Hello;
        length = sizeof( _Hello);

        // 2. play back entire audio resource
        while ( length > 0 )
            feedMP3( &p, &length);

        flushMP3();                 // flush the buffer

        // 3. repeat after a brief pause
        DelayMs(200);

    } // main loop
}
```

Listing 6.13 – main.c, Playing an MP3 message from flash

Note

Thanks to the versatility of the VS1053 decoder chip, the same code presented in this example can be used without modification to playback ANY audio resource formatted using one of the many compatible codecs, including: WAV, Ogg Vorbis, WMA, AAC and MIDI!

Playing Audio Files

The next logical step in our progression is to attempt to play back audio files directly from a micro SD card. This would clearly eliminate any concern regarding the size of the resource and competition for the limited flash program memory.

For this new project, that we will call: **Audio_Play**, we can start with a copy of the previous example project (or with the checklist introduced at the beginning of the chapter) and then re-introduce a number of library modules that we have used and developed in the previous chapters, including the following MLA modules:

- *IL9341.c (or HX8347.c)*, from the Graphics/Drivers folder
- *Primitives.c*, from the Graphics folder
- *SD-SPI.c*, from the MDD File System folder
- *FSIO.c*, from the MDD File System

Let's create a new ***main.c*** source file. In Listing 6.14, you will recognize the initialization steps for the Mikromedia board, the Graphics library and the the MP3 decoder, followed by MDD File System initialization.

> **NOTE**
>
> Since the MP3 decoder, the serial flash chip and the SD card interface use the same SPI port, it can end up being initialized multiple times. Fortunately the settings are identical among the three interfaces. Had that not been the case, we would have had to reconfigure the SPI port each time we had to alternate its use between incompatible devices.
>
> In practice we could eliminate some code duplication and share the SPI port low level access routines to save a little program memory space, but it is actually more convenient to maintain the separation of the libraries rather than customizing them for our specific use case.

Eventually we will attempt to open a media file: *SONG.MP3* for reading.
A very basic diagnostic indication is provided in this demo solely by changing the color of the LCD display according to the following code:
- *white*, at initialization and while waiting for a card to be inserted
- *red*, if the specified file name cannot be found
- *green*, when the file is found and playback is started

```c
/*
 * Project: AudioPlay
 *
 * File:    main.c
 */

#include "PICconfig.h"

#include "Graphics/Graphics.h"
#include "MDD File System/FSIO.h"
#include "uMedia.h"
#include "VS1053.h"

int main( void )
{
    FSFILE *fp;
    BYTE data[2048];
    size_t length;
    BYTE *p;

    // 1. initializations
    uMBInit();                          // initialize PPS

    InitGraph();                        // initialize Graphics
    SetColor( WHITE);
    ClearDevice();
    DisplayBacklightOn();               // white screen


    // init the MP3 Decoder
    InitMP3( 0);                        // MP3 decoder enabled
    setMP3Volume( 30, 30);

    // init file system, wait for SD card to be inserted
    while  ( FSInit() != TRUE)
    {
        DelayMs(100);
    }

    // signal card detected and mounted
    SetColor( GREEN);
    ClearDevice();          // show green screen if successful initializing

    // try to open an MP3 file
    if ( (fp = FSfopen( "SONG.MP3", "r")) == NULL)
    {
        SetColor( BRIGHTRED);
        ClearDevice();
        while(1);
    }
```

Listing 6.14 - AudioPlay, main.c, initialization

```
    // 2. Main Loop
    while( 1 )
    {
        int i;
        // 3. check if buffer ready
        if (length == 0)
        {
            MP3_DCS_Disable();

            // 4. fetch more data
            length = FSfread( data, 1, sizeof(data), fp);
            p = &data[0];

            if (length==0)           // 6. eof
            {
                flushMP3();          // flush buffer
                FSrewind( fp);       // rewind file
                DelayMs(200);        // repeat after a brief pause
                continue;
            }
        }

        // wait if codec not ready to get more data
        while ( !MP3_DREQ)
        {
            // add your task here
        }

        // 5. feed the codec
        MP3_DCS_Enable();            // select the data interface
        for( i = 0; i < 32; i++ )
        {
            if( length == 0)         // if sent all data exit
                break;

            writeMP3( *p);
            p++;
            length--;                // decrement counter
        }
    } // main loop
}
```

Listing 6.15 – AudioPlay, main.c, main loop

It is in Listing 6.15, that the core of the playback loop takes place:
A simple buffering mechanism is implemented here, using the data array, to hold a large number of samples, so that the operations of reading from the SD card (4.) and writing to the MP3 decoder (5.) can be decoupled leaving room in between for other tasks to be executed in parallel.
When the playback is completed (reaching the end of file in 2.), the MP3 decoder buffer is flushed (6.) and the file is rewound so to repeat after a brief pause.

The simplicity of this arrangement is obvious, but there are several elements of the design that might require testing and a bit of experimentation.
For example the size of the *data* buffer was chosen conveniently identical to the size of the entire decoder FIFO. This could be a sub-optimal choice as RAM is a precious resource and

in fact it could be demonstrated that a smaller buffer can serve better applications that require more "granularity", that is a more frequent switching among tasks. Eventually the specific application run time requirements will dictate which strategy to choose for optimal buffering.

Build and Run the project to verify the proper and continuous playback of the media file. A little experimentation with files encoded in different formats, compression ratios and different sample rates will empirically demonstrate the amount of workload required by the microcontroller to keep up with the streaming task and consequently the performance that can be expected for any parallel task required.

Tips and Tricks

The VS1053 can also be used as a PCM decoder to generate arbitrary waveforms by sending a WAV file header to its data interface. If the length sent in the WAV header is 0xFFFFFFFF, the VS1053 will stay in PCM mode indefinitely (or until SM CANCEL has been set). Both 8-bit linear and 16-bit linear audio is supported in mono or stereo.
A WAV header looks like this:

File Offset	Field Name	Size	Bytes	Description
0	ChunkID	4	"RIFF"	
4	ChunkSize	4	0xff 0xff 0xff 0xff	
8	Format	4	"WAVE"	
12	SubChunk1ID	4	"fmt "	
16	SubChunk1Size	4	0x10 0x0 0x0 0x0	16
20	AudioFormat	2	0x1 0x0	Linear PCM
22	NumOfChannels	2	C0 C1	1 for mono, 2 for stereo
24	SampleRate	4	S0 S1 S2 S3	0x1f40 for 8 kHz
28	ByteRate	4	R0 R1 R2 R3	0x3e80 for 8 kHz 16-bit mono
32	BlockAlign	2	A0 A1	2 for mono, 4 for stereo 16-bit
34	BitsPerSample	2	B0 B1	16 for 16-bit data
52	SubChunk2ID	4	"data"	
56	SubChunk2Size	4	0xff 0xff 0xff 0xff	Data size

<u>Table 6.3 - WAV Header</u>

The rules to calculate the four variables are as follows:
- S = sample rate in Hz, e.g. 44100 for 44.1 kHz.

- For 8-bit data B=8, and for 16-bit data B=16.
- For mono data C =1, for stereo data C =2.
- A = (C×B) /8
- R=S×A.

For example, a 44,100 Hz 16-bit stereo PCM header would read as follows:

```
0000 52 49 46 46 ff ff ff ff 57 41 56 45 66 6d 74 20  |RIFF....WAVEfmt 
0100 10 00 00 00 01 00 02 00 44 ac 00 00 10 b1 02 00  |........D.......
0200 04 00 10 00 64 61 74 61 ff ff ff ff              |....data....
```

Summary

In this chapter we have started to explore the capabilities of the VS1053, a modern audio codec available on the Mikromedia board, to augment our applications with audio capabilities. We learned how to test the device first, and later how to reproduce an audio resource from the microcontroller flash memory and eventually from a micro SD card.

Suggested Reading

- Di Jasio, "Programming 16-bit Microcontrollers in C", Newnes Elsevier
    - Chapter 15, Volare – Reproducing basic PCM Audio using a microcontroller PWM module

Online Resources

- http://www.flyingpic24.com

    You can find here links to all the resources needed for this and previous books, including code repositories, errata and updates

- **Audacity** – http://audacity.sourceforge.net,

    The home page of the Audacity open source project

Exercises

1. Add menu file/folder selection to the Audio Player

    (Hint: use the LCD terminal emulation developed in Chapter 3 and the touch screen menu module developed in Chapter 4)

2. Add automatic playback of all media available in a given folder

    (Hint: use the `FindFirst()`/`FindNext()` functions, as we did in Chapter 5 to scan through all the files in a folder)

3. Create an interrupt based playback loop, so that audio playback is executed in the background while the main application loop is in the foreground.

Chapter 7

Graphics Object Layer

It is perhaps because of the ubiquitous smart phones and the advent of the tablets era that, in recent years, the definition of an embedded control *user interface* has changed dramatically. While far from the performance (size, resolution and color) of any smartphone or tablet, the QVGA display demonstrated on the Mikromedia board is nowadays considered a minimum-requirement for any modern application. The MLA Graphics library, with its graphic primitives we have explored in the first couple of chapters, can be used to design any number of such interfaces, but starting from that level to design even the simplest graphical user interface (GUI) can be quite a challenging proposition. For this reason a significant portion of the code in the MLA Graphics library has actually been spent to offer the developer a ready to use, extensible, GUI toolkit: the *Graphics Object Library* or *GOL*. In this chapter we will explore the fundamentals of designing with the GOL and its associated rapid development tool: the *Graphics Display Designer* or *GDD*.

An Overview of GOL

The name of this Graphics library is already giving us one important clue about the key design philosophy at its core: it's *object oriented*. But, perhaps this could come as a surprise to some of you, the library is actually written entirely in plain C language and NOT in C++ or other modern object oriented language!

This is a great advantage, as it is possible to use the same source code across all three major architectures (8-bit, 16-bit and 32-bit PIC® microcontroller) using the MPLAB® XC compiler suite. It is also a limitation as there is a significant amount of redundancy and verbosity in this implementation that could have been avoided using a "true object oriented" language. Similarly a language that would support directly memory allocation and garbage collection would have simplified and perhaps optimized the use of RAM resources although, quite possibly, at the expense of portability, performance and true real time operation.

The designers of the GOL library opted instead for the use of the most basic memory management (as provided by all three XC compilers) so to maintain cross-architectural independence and maximum performance predictability. This means that when using GOL, we will have to manually set aside some RAM memory for the "heap" and, while the library will allocate the memory when new "objects" are created, it will be our responsibility to remember to free it up when the same objects won't be used anymore.

Built on top of the MLA Graphics Primitive layer, the GOL library benefits from the extreme portability of its code across many display manufacturers, control interfaces and microcontroller interface peripherals options. If you have read the first three chapters of this book, you will have an immediate understanding of how any of the GOL examples available can be ported to the Mikromedia board with the simple customization (re-use) of the *GraphicsConfig.h* and *HardwareProfile.h* modules.

Widgets and Objects

If you have ever designed an application for a personal computer in the last 10 years, be it Windows, Linux or OS X, anything beyond the most basic command line will have had you exposed to some of the fundamental building blocks of GUI design: the *widgets*.

Here is a list of the basic GOL widgets available as of this writing:

1. Button
2. Slider
3. StaticText
4. TextEntry
5. Picture
6. CheckBox
7. EditBox
8. ListBox
9. ProgressBar
10. RadioButton
11. Window

There are then additional and more complex widgets such as:

12. Grid
13. GroupBox
14. RoundDial
15. Chart
16. Palette
17. Meter
18. DigitalMeter

19. AnalogClock

These elemental components of the GOL design are offered as individual source files and, like Lego blocks, can be assembled to form any desired user interface.

They receive inputs and communicate with the main application and among each other via a *message passing* mechanism. By carefully handling the messages they send and receive it is possible to combine them to form complex new widgets. Eventually, they can be hand customized to satisfy very specific application needs, creating entirely new widgets and expanding the expressive power of the GOL.

Blocking vs. Non-Blocking

While this is not the case of the display controller used by the Mikromedia boards (HX8347 or IL9341), some of the display controllers supported by the MLA Graphics Primitives have powerful hardware accelerators on board which can take much of the typical workload of the microcontroller off during rendering of complex objects.

For example, on a basic display controller, the act of filling a large area of the screen with a new background color can require the PIC microcontroller to access the display hundreds of thousands of times. A hardware accelerator can perform the same operation after receiving a single command requiring only a handful of bytes of data to be transferred from the PIC microcontroller and in a much shorter amount of time by working directly on the display board and memory array.

It would be wasteful though to have the PIC microcontroller waiting for the hardware accelerator to complete each operation when it could instead make better use of the time by tending to other more pressing activities.

The MLA GOL library can be configured (defining the USE_NONBLOCKING_CONFIG symbol in *GraphicsConfig.h*) so to split drawing and updating of widgets in a sequence of elemental steps that can be executed asynchronously. In practice, if rendering a button widget on the screen can be imagined as a sequence of four steps:

1. Clearing the background
2. Drawing the shaded edges
3. Drawing an image inside it
4. Rendering its text

In non-blocking mode, the drawing function (GOLDraw() as we will learn shortly) will return control immediately after launching the first step, in fact without even waiting for it to be completed.

A state machine, internal to the button widget, will ensure that all steps are *eventually* completed through the following rounds in the main application loop.

This means that fine-grained cooperative multitasking is possible with the MLA GOL library and it can be used to advantage when a graphics accelerator is available.

At the same time, proper non-blocking application design is more complex and can require additional code to ensure the necessary synchronization when the application demands it. In the following examples in this chapter we will refer exclusively to the use of the GOL library in blocking mode.

> **NOTE**
>
> While designing the GOL library around an RTOS would have achieved similar or likely better overall performance by providing all the tools to handle multitasking, once more the choice of the designers has been in favor of cross-platform portability and simplicity.

Object States

Regardless of the use of blocking or non-blocking mode and the related state machine, all widgets carry additional state information that is used to control their behavior.

Property States, or simply *Properties,* are used to keep track of the widget appearance, such as for example:

- OBJ_DISABLED, which determines if it should completely ignore any incoming message
- OBJ_FOCUSED, which determines if the widget should receive user input

Drawing States are similarly used as flags to indicate that specific actions are required such as:

- OBJ_HIDE, which determines if the object should be visible at all at any given point in time
- OBJ_DRAW, which indicates whether it should be repainted as part or all of it got exposed to the user view

Properties and drawing state of an object are most often modified automatically by the GOL library during creation and normal operation but can also be modified manually by the user as required.

About Style

Some describe the GOL widgets appearance as somewhat reminiscent of the early Windows 3.0 look, with its 3-D shading/embossing effects if only more recently updated with a touch of gradient fills. So it is not exactly IOS 7, parallax adjusted and heavily animated stuff, but it can be a good value nonetheless for most embedded applications. It is, of course, a compromise between pleasing the eyes and keeping resources and processor usage low while supporting the largest range of architectures (from a PIC18 to the latest PIC32). Naturally, having the source code readily available means that we can customize and enhance the look at our heart's content if we determine that our applications deserve more and we can afford to spend the extra resources and processor cycles.

When the GOL documentation mentions *style* though, it is to refer to the choice of a set of colors and fonts that can be shared among (all or a subset of) widgets in a given application.

> **NOTE**
>
> While it is absolutely possible to associate each widget on the screen to a different style scheme, the result can be quite psychedelic. It is almost always better to exercise extreme restrain in the matter. Choose carefully a set of colors, a font size and style, and use it consistently and deliberately across your application interface pages.

A *GOL style scheme* is a *C struct* (defined in *GOL.h*) containing a few integers (16-bit RGB values) describing a pair of widgets body colors, text colors and a pointer to a font resource. The actual usage of the colors in the style scheme is dependent on each specific widget design. In other words there is no fast and simple rule to tell what `TextColor0` rather then `TextColor1` will be used for unless you actually check the documentation of a specific widget. An example is given in Figure 7.1 for a basic Button widget.

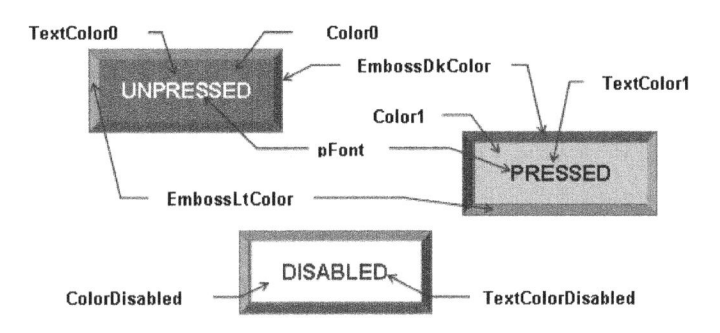

Figure 7.1 – Style Scheme usage of a Button Widget

Creating a Widget

As expected, the GOL library has its own initialization function, `GOLInit()`, that in turns takes care of calling the primitive graphics library initialization function and creates all the basic infrastructure necessary for the graphical user interface creation.

Once it's called, we can start *instantiating* new widgets right away. That is object oriented lingo for "creating an instance of" any specific kind of object.

Say we want to add a button to the top left of the display, we can use:

```
pButton = BtnCreate(
        10,                // object's ID
        10, 10, 110, 60,   // left, top, right, bottom coordinates
        0,                 // radius of rounded edge, use pointy edges
        BTN_DRAW,          // draw the object after creation
        NULL,              // no bitmap used
        "Done",            // text to be shown inside the button
        NULL               // use default style scheme
        );
```

Listing 7.1 - Button instantiation

All *widget-create* functions look very much alike. The common sequence of parameters includes a basic set of integers as follows:

- *Object ID*, this is just an integer value that will come handy later to refer to this specific widget. Naturally this number better be unique, at least within the current "page", that is the specific group of objects in use at any given point in time. But, other than that, it can be assigned according to any desired rule or personal habit: strictly sequential (1,2,3...), in increments of ten (10, 20, ...), grouped in hundreds (all buttons are 1xx, checkboxes 2xx, etc.) there are reasons to justify any of the above techniques and more. Using a `#define` directive we can associate a descriptive name to what would otherwise look like an unremarkable literal.

- *Left, Top, Right and Bottom*, are the absolute coordinates defining the position and size of a widget. Values assigned here will be used to define the *outside* boundaries of the widget, which, in turn, will define the *internal* area used to draw the widget content. Each widget takes these coordinates very seriously, there are no checks made to ensure that the space given is sufficient to fit the desired text and images. On the contrary, the content is *clipped* so that in the worst case, the widget will look cramped (or just plain ugly) but it will not corrupt neighboring widgets. More importantly the coordinates are going to be used to determine the position and size of the touch sensitive area of the widget. It is easy to feel the pressure to pack as many information and controls on any given screenful. Unless we expect users to carry a stylus with them at all times, we must ensure that enough room is

given for a typical (adult) finger to clearly reach and operate all the touch elements (buttons, sliders...)
- *State*, gives us the option to specify properties and initial drawing states of a widget. It is typical to request it to be drawn right away for example, but we can easily imagine how it could be useful to keep some interface elements hidden initially, only to be enabled and displayed at a later time upon a particular event taking place.

These are followed by up to three pointers:
- *Image pointer*, provides a pointer to the resource if an image is added to the representation of the widget. Since image resources can be in flash or external, the parameter is defined as a `void*` to avoid the compiler worrying about pointer type matching. Passing a `NULL` pointer will default to no image.
- *Text pointer,* provides an intial value if a text element is required/displayed by the widget. Note that this can be a string variable (RAM), or a string literal (flash). A `NULL` value, will default to no text being displayed.
- *Style Scheme pointer,* is the link to the style scheme structure presented above, which defines the color scheme and the font resource to be used with the widget. A `NULL` value here will default to a predefined *GOL Default Style Scheme* which is not particularly attractive, but convenient and known to work.

Additional parameters are occasionally added (unfortunately not always in a consistent position) to the create functions of specific widget types. In our previous example (see Listing 7.1), the *Radius* parameter, was passed to the `BtnCreate()` function to specify an optional styling element after the position and size.

That might seem like a lot of parameters to pass to a function, but we must consider that this is something that needs to be done just once per view/page. From now on, access to the Button widget will be obtained by using either the pointer returned, `pButton` in our example or, a bit more indirectly, via the Object ID value.

In fact most beginners are confused as to which one of the two to use and when: the object pointer or the object ID?

While some of the functions in the GOL API do offer both options, for example: `GOLDeleteObject()` and `GOLDeleteObjectByID()`, most functions only accept object pointers. Still, it can be convenient to refer to the object ID when handling messages among widgets to make the code more readable (and possibly save a few RAM locations).

Otherwise it is quite simple to obtain an ID (my_id) from a given object pointer (my_p):

    my_id = GetObjID(my_p);

And, vice versa, obtaining an object pointer (my_p) from a given object ID (my_id):

    my_p = GOLFindObject(my_id);

Notice that while GetObjID() is a simple macro, producing a pointer de-reference taking a very small (fixed) amount of processor cycles, GOLFindObject() performs a linear search through the list of known objects, which takes a much larger amount of cycles.

Messaging

Messaging is the fundamental mechanism used to supply input to widgets and for communication among widgets. A *message* is a small data structure defined in *GOL.h* as follows:

```
typedef struct {
    BYTE type;
    BYTE uiEvent;
    SHORT param1;
    SHORT param2;
} GOL_MSG;
```

The type field can be one of the following values:

- TYPE_UNKNOWN
- TYPE_KEYBOARD
- TYPE_TOUCHSCREEN
- TYPE_MOUSE
- TYPE_TIMER
- TYPE_SYSTEM

which, for all practical purposes, can be restricted to the sole TYPE_TOUCHSCREEN when used on the Mikromedia board unless we add a USB mouse or keyboard.

Each type of input device then defines a number of possible events that it might want to notify to the GOL widgets. In the touchscreen case those are:

- EVENT_INVALID
- EVENT_MOVE
- EVENT_PRESS
- EVENT_STILLPRESS
- EVENT_RELEASE

For each event type a precise use of the `param1` and `param2` fields is dictated. For all the touchscreen events these two fields are simply used to contain the x and y coordinates on the screen. The responsibility of creating such input messages is given to a board specific support modules which, in the Mikromedia board case, is the *touchscreen.c* module via the `TouchGetMsg()` function:

```c
void TouchGetMsg(GOL_MSG *pMsg)
{
    static SHORT    prevX = -1;
    static SHORT    prevY = -1;
    SHORT           x, y;

    x = TouchGetX();
    y = TouchGetY();
    pMsg->type = TYPE_TOUCHSCREEN;
    pMsg->uiEvent = EVENT_INVALID;

    if ( (x == -1) || (y == -1))
    {
        y = -1;
        x = -1;
    }
    if ( (prevX == x) && (prevY == y) && (x != -1) && (y != -1))
    {
        pMsg->uiEvent = EVENT_STILLPRESS;
        pMsg->param1 = x;
        pMsg->param2 = y;
        return;
    }
    if ( (prevX != -1) || (prevY != -1))
    {
        if((x != -1) && (y != -1))
        {   // Move
            pMsg->uiEvent = EVENT_MOVE;
        }
        else
        {   // Released
            pMsg->uiEvent = EVENT_RELEASE;
            pMsg->param1 = prevX;
            pMsg->param2 = prevY;
            prevX = x;
            prevY = y;
            return;
        }
    }
    else
    {
        if ( (x != -1) && (y != -1))
        {   // Pressed
            pMsg->uiEvent = EVENT_PRESS;
        }

        else {   // No message
            pMsg->uiEvent = EVENT_INVALID;
        }
    }

    pMsg->param1 = x;
    pMsg->param2 = y;
    prevX = x;
    prevY = y;
}
```

Listing 7.2 – TouchGetMsg() function, *touchscreen.c* module (continued)

Once an input event message is produced, it must be passed sequentially to all widgets currently defined and active (by calling the `GOLMsg()` function) to verify if it is affecting the respective portion of screen surface .

This way each widget is given the option to handle the event by taking immediate action and additionally translating the input message generating a new *translated-message*.

> **NOTE**
>
> The re-use of the term "message" here can be the source of great confusion.

A translated-message is NOT a data structure of the original `GOL_MSG` type or a modification of it, but rather a single *integer* value part of a short list of possible values (`TRANS_MSG` enumeration) defined in *GOL.h* such as:

```
typedef enum {
    OBJ_MSG_INVALID = 0,
    CB_MSG_CHECKED,
    CB_MSG_UNCHECKED,
    ...
    BTN_MSG_PRESSED,
    BTN_MSG_STILLPRESSED,
    BTN_MSG_RELEASED,
    BTN_MSG_CANCELPRESS,
    ...
    OBJ_MSG_PASSIVE
} TRANS_MSG;
```

Listing 7.3 - *Translated* messages enumeration

For example, pressing a button (touching the screen in the area covered by the Button type widget), will change automatically its appearance on the screen, but will also return a *translated message* value: `BTN_MSG_PRESSED`.

The translated message value, together with the original input message and a pointer to the object that accepted it, will be now passed (by `GOLMsg()`) to the all important, user defined `GOLMsgCallback()` function.

> **NOTE**
>
> `GOLMsgCallback()` is where we get the precious opportunity to interact with the user interface. This is where we get to *intercept* messages; hence the true application logic lives here!

The Active Object List

We have already mentioned a couple of times how a view/page of the user interface is going to be composed by a number of widgets. So you will be probably curious to know now how a *page* (from now on I will stick to this term) is assembled and maintained.

The fundamental (if only hidden) data structure here is a very simple linked list of widgets known as the *Active Objects List*.

The root (head) link is a pointer known as `_pGOLObjects`.

But it would very inappropriate to use it directly; in fact there are very few reasons to do so as most GOL library functions refer to it *implicitly*:

- When we initialize the library, `GOLInit()` properly initializes it to an empty list

- When we create a new widget, using `BtnCreate()` for example, the new widget data structure is dynamically allocated and automatically linked (appended) to the list

- When we search for an object by ID, the `GOLFindObj()` function performs a linear scan of the entire list looking for a match

- When an input message is generated by the touch screen module, the `GOLMsg()` function is used to feed it to the active widgets by scanning the list

- When drawing and updating all the widget present, the `GOLDraw()` function invokes the draw method for each widget in the list

Drawing

We saved for last what is perhaps the most obvious of all the GOL library tasks: drawing widgets on the screen. This is performed by the `GOLDraw()` function, and it is our job to invoke it repeatedly in what is the main loop of the user interface.

The `GOLDraw()` function is actually quite a complex state machine. As we have seen in the introduction, it can operate in Blocking and Non-Blocking mode to increase the granularity of each widget rendering task. It is also designed to allow additional flexibility by providing a callback mechanism of its own. In fact the user is *always* required to provide a `GOLDrawCallback()` function even if this means simply providing an empty shell.

The callback function is called automatically only once `GOLDraw()` has completed the update of all widgets in the active list so that it is safe to use any of the graphics primitives (altering font and color selections as needed) without fear of interfering with the main drawing task.

Typical use of the `GOLDrawCallback()` function is to provide custom "visual" extensions to existing widgets without requiring necessarily the creation of a complete new widget *type*. The latter option is always available, and well documented (see links to available application notes in appendix) but necessarily more complex as it requires the manual modification of most/all core modules of the GOL library.

The return value of the callback function is supposed to indicate whether the custom drawing task was completed or not. In case of a positive (`True, 1`) response, the regular `GOLDraw()` activity is immediately resumed at the next call through the main loop. But if a False (0) value is returned, the update of the active objects list is suspended, and the callback function will be called immediately upon the next call of the drawing function in the main loop.

This is designed so that the callback function can fragment its work should it require long periods of time that could otherwise affect the overall application responsiveness.

A First GOL (Application)

As a first example of usage of the GOL library module, we will create the most basic user interface composed of a single button centered in an otherwise empty and dark screen.

Preparation

As in all previous chapters, let's start by creating a new MPLAB X project.
The pre-requisite elements are clearly all the same supporting MLA modules for graphics primitives, touchscreen and default font usage. These will, in their turn, require the addition of the Serial Flash module to store and retrieve the touch calibration values, including the SPI driver support and the M25P80 memory module.
Here the simple steps:

1. Use the **New Project Wizard** to create a new project inside the working directory (Mikromedia), let's call it: "**GOL_Simple**"

2. Add a new logical folder to the Source Files folder, called **MLA**

3. Add the following items to the folder:
   - **IL9341.c or HX8347.c**, the display controller driver
   - **Primitive.c**, the primitive layer of the graphics library
   - **TimeDelay.c**, a few basic timing (blocking) functions used by the driver
   - **GOLFontDefault.c**, a default font resource
   - **GOLSchemeDefault.c**, a default color scheme
   - **GOL.c**, the main GOL module
   - **Button.c**, the button widget source

4. Add a new logical folder to the Source Files folder, called **uMB**

5. Add the following items to the folder:
   - **drv_spi.c**, the SPI driver for access to the serial flash
   - **M25P80.c**, the serial flash driver
   - **TouchScreen.c**, the touchscreen main module
   - **TouchScreenResistive.c**, the resistive touchscreen specific module
   - **uMedia.c**, the basic Mikromedia board initialization functions

6. In the *XC16-gcc* settings, configure the compiler **C include dirs** to contain:

- **. (dot),** the current project directory for MLA to reach our configuration files
- **../Microchip/Include,** for our source files to reach inside the MLA
- **../uMB, f**or project source files and the MLA to access the Hardware Profile and other resources specific to the Mikromedia board.

7. In the *XC16-ld* settings, set a value for the Heap Size: **2048**

8. Create a new **main.c** file using the **New File wizard** and the embedded template or your own customized version

```c
/*
 * Project: 7-GOL_Simple
 *
 * File:    main.c
 *
 */
#include "PICconfig.h"
#include "HardwareProfile.h"
#include "uMedia.h"

#include <Graphics/GOL.h>
#include <TouchScreen.h>
#include <M25P80.h>

int main( void)
{
    GOL_MSG msg;

    // 1. initializations
    uMBInit();
    TickInit( 1);
    TouchInit( NVMWrite, NVMRead, NVMSectorErase, NULL);

    GOLInit(); DisplayBacklightOn();

    BtnCreate(  1,                          // unique ID
                100, 80, 220, 160,          // position and size
                0,                          // radius
                BTN_DRAW,                   // state
                NULL,                       // no bitmap
                "Done",                     // text
                NULL                        // default style scheme
            );

    // 2. main loop
    while( 1)
    {
        if ( GOLDraw())                     // if done drawing the objects
        {
            TouchGetMsg( &msg);             // generate a messsage if touched
            GOLMsg( &msg);                  // process the message
        }

    } // main interface loop

} // main
```

Listing 7.4 – First GOL example, main() function

Graphics Object Layer - 193

In the new *main.c* file we can start now adding the usual include directives and the `main()` function as in Listing 7.4:

In the initialization section (1.), the only novelty is represented by the call to `GOLInit()` followed by the screen backlight activation.

In the main loop (2.), we see the standard sequence of calls to `GOLDraw()` and `GOLMsg()` as required for the proper functioning of any GOL interface.

Two additional functions must be added to the *main.c* file (as per Listing 7.5) to provide the mandatory although empty (in this case) callback hooks:

```
WORD GOLMsgCallback( WORD objMsg, OBJ_HEADER* pObj, GOL_MSG* pMsg)
{
    return 1;
} // GOL Msg Callback

WORD GOLDrawCallback()
{
    return 1;
} // GOL Draw Callback
```

Listing 7.5 – First GOL example, Callback functions

As a final touch, let's make sure that we have a copy of the *GraphicsConfig.h* file in our project directory and that it has the GOL and button widget "switches" both enabled (uncommented):

```
#define USE_GOL          // Enable Graphics Object Layer.
#define USE_BUTTON       // Enable Button Object.
```

GOLSimple Testing

After saving the project and building it successfully, you should be able to obtain a screenshot similar to what captured in Figure 7.2:

Figure 7.2 – GOLSimple project, screen capture

Admittedly this is not a very exciting application, but it can give us a feel for the kind of look and behavior we are going to obtain from a basic button widget.
The standard style provides a pale blue background for the button and the text does change color as the button is touched to provide a 3D illusion of downward motion.

A Slider Example

Although the button in our first example did say: "Done", nothing much was happening besides its image being updated on screen. For the next example we will switch instead to a *Slider* widget and we will attempt to connect the position of the "thumb" or "cursor" to the backlight of the LCD display.

We can start modifying the previous project or, we'd better make a copy of the entire project (directory) using the Project context menu and selecting the **Copy** function. Afterwards we can rename the entire project: **"GOL_Slider"**.

First, we will need to find a way to adjust the luminosity of the screen similarly to what we did in the solution to chapter 2's exercise. Let's start by defining a `BacklightInit()` function (see Listing 7.6) that will configure and connect the *OutputCompare* module `OC1` in PWM mode to output on the same `RD2` pin which is in control of the LED display backlight.

We can set the period of the PWM output to a fixed value of 240Hz, high enough to eliminate any visible flicker, and conveniently providing 16-bit resolution.

```
void BacklightInit()
{
    // configure OC1 block to generate a PWM signal
    OC1CON1bits.OCTSEL = 0x7;        // use peripheral clock (16Mhz)
    OC1CON1bits.OCM = 0x6;           // edge aligned PWM
    OC1CON2 = 0;                     // No SYnchronizations required
    OC1R  = 0xfffe;                  // start off with max value
    OC1RS = 0xffff;                  // period ~240Hz (16MHz/65.536)
    PPSOutput( PPS_RP23, PPS_OC1);   // OC1 =RP23 D2/pin 77
} // Backlight Init
```
Listing 7.6 - BacklightInit() function

Notice how the PPSOutput() function (provided by the PPS peripheral library module) is used to map the OC1 output to the correct pin. This requires that the **PPS.h** file be included at the top of our main file, but also requires that the PPS module be left "unlocked" after power up and any prior configuration (as performed in **uMedia.c**).

After the GOLInit() call, let's now make sure to call the BacklightInit() function in place of the standard DisplayBacklighOn() macro.

```
// 1. initializations
uMBInit();
TickInit( 1);
TouchInit( NVMWrite, NVMRead, NVMSectorErase, NULL);
GOLInit();
BacklightInit();                // init to full brightness
```

By controlling the PWM duty cycle, we will be able now to dim the backlight with more than sufficient resolution, too much in fact!
For simplicity we will define the BacklightSet() function to take a parameter between 0-255 and scale it up to the full duty cycle range by shifting it left by 8 positions.

```
void BacklightSet( unsigned i)
{
    OC1R = i<<8;
}
```

Listing 7.7 - BacklightSet() function

Both functions can be added to the **uMedia.c** module (remember to put the appropriate prototypes in the **uMedia.h** header file as well) for later re-use.

Creating and initializing a Slider widget is easily accomplished during the program initialization (within *main.c*) with the following function call:

```
SldCreate( 1,                      // unique ID
           20, 100, 300, 140,      // position and size
           SLD_DRAW,               // state
           250,                    // 0..range
           25,                     // resolution step
           250,                    // initial position (100%)
           NULL                    // default style scheme
        );
```

Listing 7.8 – SldCreate() example

This function resembles the `BtnCreate()` function it replaces except for the last three parameters that are clearly slider specific:

- The *range* parameter is defining how the position of the slider is scaled to produce an output value. The left most position (for horizontal sliders, or bottom for vertical ones) is always producing a 0 (zero) output. The right most (or top) position will return the *range* value. Values in between will be linearly interpolated. In our example the output value will be ranging from 0 to 250.

- The *resolution* parameter, (referred to as *page* in the official documentation) defines the granularity of the thumb movement and output value scaling. In our example this will be in increments of 25, which means the output value will be moving in discrete steps: 0, 25, 50, 75,.., 250

- Finally the *position* parameter, defines the initial position of the thumb upon widget creation and when first drawn on the screen.

Creating a Window

While we are at it, we will take the opportunity to introduce another widget: the "Window". Although the name might have you thinking otherwise, this is actually a very basic *container* object. It can be used to group a number of smaller widgets or, as is the case in our example, simply to place a banner on top of the screen and to control the background color.

This is accomplished by defining a window of the size of the entire screen (using the `GetMaxX()` and `GetMaxY()` functions) and by defining the background and banner colors in a custom color scheme.

```
wndCreate( 2,                          // unique ID
           0, 0, GetMaxX(), GetMaxY(),
           WND_DRAW_TITLE | WND_TITLECENTER,
           NULL,                       // icon
           "Slider Demo",              // window title
           myStyle                     // default style scheme
         );
```
Listing 7.9 - WndCreate() example

Once more the `WndCreate()` function has a small number of custom parameters (the last three) we need to carefully provide:

- A pointer to an *icon* object, which would be displayed on the title bar (not used in our example)
- The *window title,* provided as a string. It can be left aligned (default) or centered by adding the property `WND_TITLECENTER` to the state parameter.

Defining A Color Scheme

A *color scheme* can be provided as the last parameter in all the widget-create functions. If a custom one needs to be defined, it must be done before the widget create function call.
In our example we will define a window with a black background, without visible embossing of the frame and a bright yellow title bar.

If `myStyle` is defined as a pointer to a `GOL_SCHEME` object:

```
GOL_SCHEME *myStyle;
```

According to Figure 7.3 (from the GOL User Guide) we will require the following scheme definition:

```
myStyle = GOLCreateScheme();
myStyle->CommonBkColor = BLACK;
myStyle->Color0 = YELLOW;
myStyle->EmbossDkColor = BLACK;
myStyle->EmbossLtColor = BLACK;
```

Listing 7.10 - Defining a new color scheme

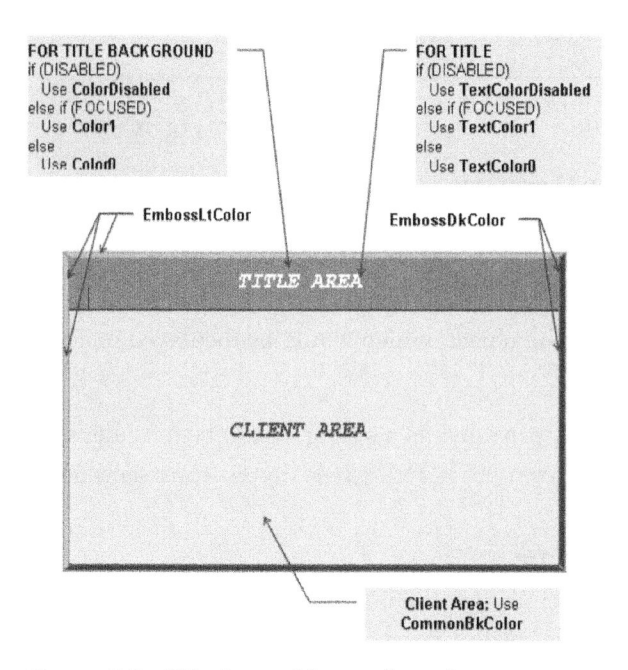

Figure 7.3 – Window widget color scheme usage

The `GOLCreateScheme()` function is used first to allocate memory for a new object of the color scheme kind, which is later filled with default values derived from a template contained in the **GOLDefaultColorScheme.c** module. Eventually only a few of the color scheme properties needed to be modified (see Listing 7.10).

Notice how this is the most common way of defining a color scheme, although not the only one possible.

GOLMsgCallback, Light, Action!

To complete the new Slider project we can now focus on the "action", which as we have seen in the introduction, is strategically performed by intercepting the right message in the `GOLMsgCallback()` function.

```
WORD GOLMsgCallback( WORD objMsg, OBJ_HEADER* pObj, GOL_MSG* pMsg)
{
    if( pObj->ID == 1)              // intercept messages from the slider
    {
        // update the screen backlight
        BacklightSet( SldGetPos( pObj));
    }
}
```

Listing 7.11- Intercepting a message from a widget

Of the three parameters passed to the callback function (Listing 7.11) we will make immediate use of the second one, the object pointer (pObj), to capture messages incoming from the slider.

Since we did not save the pointer to the slider object when creating it with the initial call to SldCreate(), we can still easily identify it by comparing the ID field (assigned during creation).

Once identified, we can obtain the slider thumb current position, as updated after the last touch event, by calling the SldGetPos() function.

Finally the returned integer can be passed to the BacklightSet() function to update the display luminosity.

Let's build the project and program the Mikromedia board with the new example code. If successful, we should be able to get an interface similar to Figure 7.4.

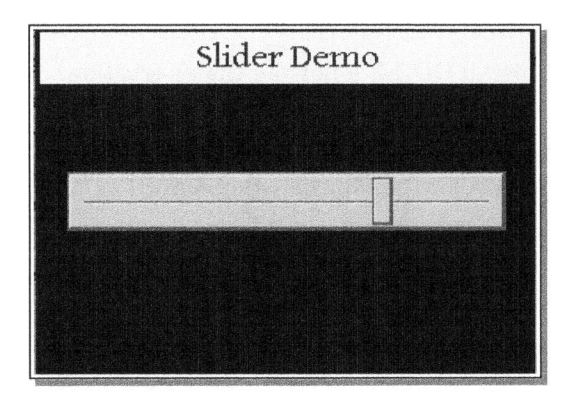

Figure 7.4 – GOL_Slider Demo , screen capture

Notice that when the slider is moved all the way to the left, the display image might seem to disappear into the dark background. Obviously the logic and touch functionality of the user interface are working regardless of the display luminous output.

NOTE

As an exercise, it would be a good idea to modify the application to provide a safe offset to guarantee a minimum usable amount of backlight is always provided.

Graphics Display Designer

When more than a handful of screens need to be assembled and the number of widgets and their interactions start growing rapidly, the work of defining each widget position and optimal size, pixel by pixel, becomes extremely time consuming and tedious. For this reason the developers of the Graphic library have devised a tool, known as the GDD X or Graphic Display Designer (for MPLAB X), that allows to assemble rapidly and efficiently all the UI elements of an application through a GUI.
That means that a graphical user interface is used to create graphical user interfaces!

GDD X is a cross platform application that can be downloaded for free from Microchip web site and installed next to MPLAB X IDE. It can be invoked from inside the IDE as a registered plug-in or as a stand-alone application.

GDD X Project Preparation

First let's download and install GDD X following the instructions appropriate for our operating system of choice.
Next, let's restart MPLAB X (to activate the new plug in) and let's create a new demo project with the usual settings for the Mikromedia board.
We will call it "**GDD_Slider**" to differentiate it from the previous examples where we used the GOL library directly.

> **IMPORTANT NOTE**
>
> The GDD X tool assumes a specific hierarchy of folders to be in place, specifically they are the *Microchip* and *Board Support Package* folders. Both are expected to be found inside the working directory!
>
> Although beginning in Chapter 2 we started copying selected contents of the *Board Support Package* folder from the MLA installation (picking only the Mikromedia related ones) into the *uMB* folder, the GDD X project wizard won't be able to link all the files required into the (automatically generated) new project structure unless it finds the *original* folder name and *all* its contents.

From inside MPLAB X Tools menu, select the **Embedded** sub-menu and eventually you will be able to launch the "**Graphics Display Designer X**" as a plug-in.
Note that this is the preferred method as the tool will now be aware of the general MPLAB X project settings and folders structure.

Graphics Object Layer - 201

From here we can start a new GDD X project by selecting the **New Project** option in the GDD X Project menu or by clicking on the New Project icon in the main GDD X toolbar.

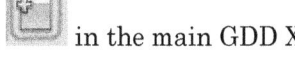

If you have familiarized with the MPLAB X IDE toolbar, you will find that the GDD X user interface is very similar and shares many of the same icons and buttons with predictable functionality and behavior.

The simple wizard will guide us through the definition of the project basic elements, in three quick steps:

- Enter the project name, type: **Slider**

- Enter the size of the screen and color resolution, select: **320x240** with **16-bit** (to match the Mikromedia display capabilities)

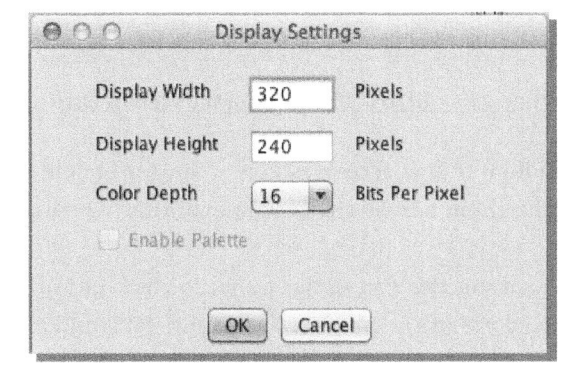

Figure 7.5 - Display Settings, dialog box

- Enter the name of a first user interface page: **SliderWindow**

GDD X will generate a new sub-folder named *Slider*, inside the *GDD_Slider.X* folder previously created by MPLAB X.

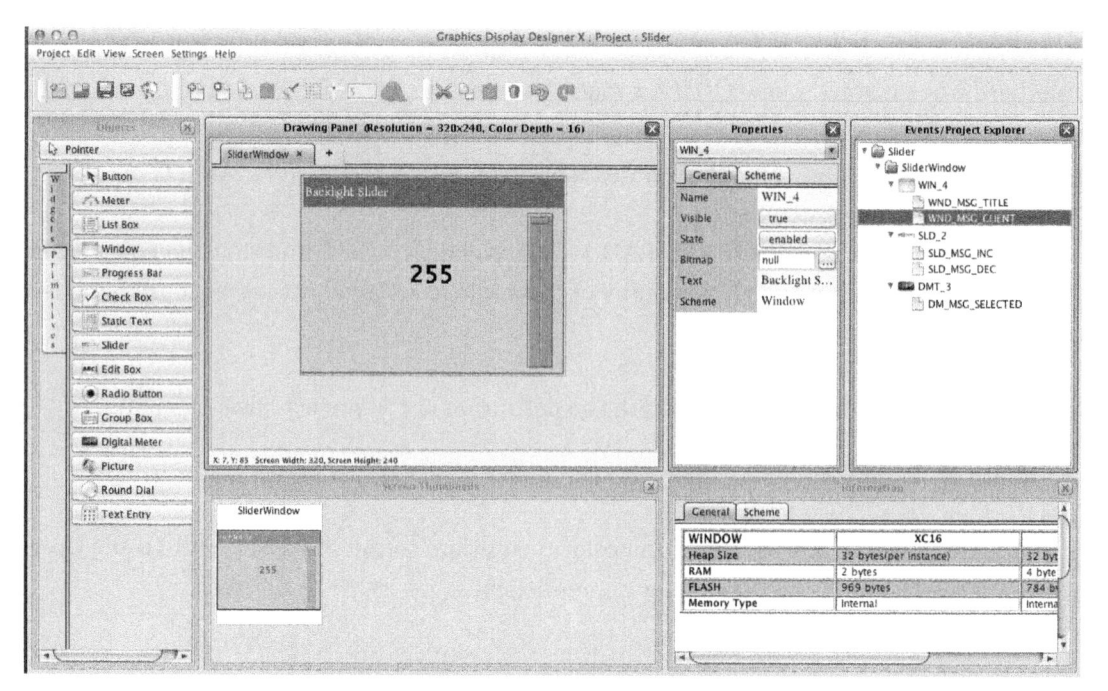

Figure 7.6 – GDD X Slider project screen capture

Adding widgets to the SliderWindow screen is now a matter of selecting them from the left *Objects* panel and dragging them across the screen eventually positioning and sizing them as desired.

Widget properties can be set via the *Properties* panel to the right of the screen.

Similarly to what we have done in the previous example, let's add a window title using the Window widget.

1. Select the **Window** widget from the left panel and place it on the screen

    - In the property pane, set the *name* to: **WIN_1**

    - Ensure it is *visible*: "**True**" and the *state* is: "**Enabled**"

    - Enter the title: "**Backlight Slider**"

2. Select the **Slider** widget from the left panel and place it on the screen

    - In the property pane, set the *name* to: **SLD_2**

    - Ensure, it is *visible*: "**True**" and the state is: "**Enabled**"

    - Set the *Orientation* to "**Vertical**", then reposition and resize to fit the right edge of the screen

- Set the *range* (255), *page size* (25) and *position* (255) as we did in the previous example

3. Select the **Digital Meter** widget from the left panel and place it in the center of the screen

    - In the property pane, set the *name* to: **DMT_3**
    - Ensure, it is *visible*: "**True**" and the *state* is: "**Enabled**"
    - Set the *Align* property to "**Center**"
    - Set *DotPos* to **0** (no decimal digits)
    - And finally set the *number of digits* to **3** (total number of digits)

At this point we are ready to save the project as defined and initiate the automatic code generation process by selecting **Project>Generate Code**!

Understanding GDD X Code

While keeping the GDD X plug-in still open we can switch back to MPLAB X (use Alt-TAB or CMD-TAB as required by your operating system of choice) and start inspecting the project structure and the code produced by looking at the **Files** window (see Figure 7.7).

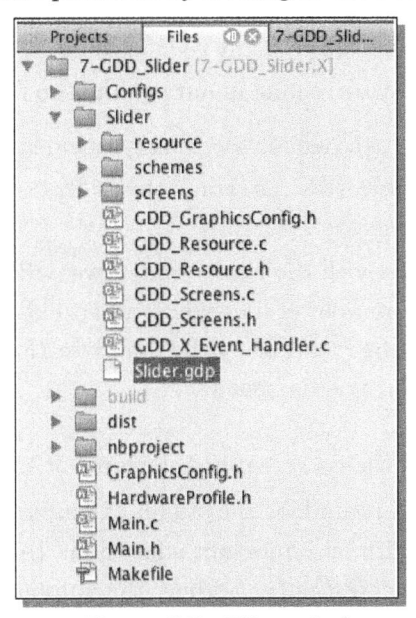

Figure 7.7 – Files window

First of all, we notice that GDD X has created a new folder within the project folder where all the GUI elements are neatly collected. This has the name of the (GDD X) project we assigned during the very first step (Slider).

Inside it we find the *Slider.gdp* file that encodes (in XML) the top-level description of the project.

We find also three additional folders that contain elements necessary for GDD to generate and configure all the screens and widgets you selected in the current project:

- *resource*, where fonts and image (.bmp) files are stored
- *scheme*, where color schemes used are stored (.xml files)
- *screen*, where each screen created is stored in a separate folder

These folders contain only the abstract representation of the GDD X project, the code we *generated* during the last step is instead found in three .c files (and their headers) present in the same *Slider* folder:

- *GDD_Resource.c*, contains the fonts and image elements ready to be packed into the microcontroller flash memory
- *GDD_Screens.c*, contains all the code required to build the widget lists for each page and to switch between pages
- *GDD_X_Event_Handler.c*, contains any code produced by actions we might have encoded using the GDD X tool (none in our example so far)
- *GDD_GraphicsConfig.h*, is a header file that is included in the default *GraphicsConfig.h* to enable (`#define`) only the widgets that have been used in the project.

Back at the top level, side by side with the *Slider* folder, we will also find a folder named: *Configs*. Inside it we will find a number of files whose names all follow the pattern HWP_*.h. Only one of them will be included by the template *HardwareProfile.h* so to customize the project for the Mikromedia board.

Finally, at the top level we find the two essential elements of each MLA project:

- *GraphicsConfig.h*, which resembles the standard configuration file we used all along in this book, but with an important addendum. In the last line it takes care to include the *GDD_GraphicsConfig.h* where the automatic *customization* takes place. Note that any changes performed by GDD X at this point are going to override previous settings

- *HardwareProfile.h*, is in reality a large switchboard that, based on the processor selected when creating the MPLAB X project, allows us to choose the hardware profile that is most appropriate for the target board. In our case, the selection of the PIC24FJ256GB110 restricts the number of known boards to a single possible choice, conveniently enough this is the Mikromedia board, therefore including the *HWP_MIKRO_8PMP.h* file from the *Configs* folder.

NOTE
> Other processor model choices might allow multiple hardware platforms to be selected. In that case we would have to manually un-comment the desired one.

And last but not least:
- *main.c*, a pretty complex "skeleton" of code that brings all the application elements together, providing all hardware modules initialization (Ports, NVM, Timing, Graphics and Touchscreen) and the main loop of a standard GOL application.

Unless we want to re-target the project for a new and custom board, *main.c* is the only file that will require our attention. We can leave all the rest to GDD X and its sophisticated selection mechanisms!

You might be tempted to remove large portions of *main.c* that are conditionally compiled when *other* processor models are selected. While this could help reduce the clutter, I do not recommend doing it this early in the process, not at least until we have a better understanding of each part of it and the somewhat complex interactions among the many symbols defined. MPLAB X color coding mechanism will quickly help us spot which portions are active and which ones are not (grayed out).

A final look at the **Project** window of MPLAB X, will reveal how all the possible GOL modules have been automatically added by GDD X to the project.
This includes all the display drivers, board support modules and primitives too!

Do not worry though, this does not mean that the application is going to be bloated by carrying unnecessary code. The MPLAB XC16 linker will take care to include in the final executable image only those modules that contain functions that have been actually used. This is just a trick employed by the GDD X project generator to simplify its operation, although it sure makes the project window look scary.

Adding User Code to a GDD X Generated Project

Opening *main.c* in the MPLAB X editor will bring to our eyes the reality of a very busy source file. This can be intimidating to say the least, but we can quickly separate some key sections of code and, once we understand their purpose, we can start customizing and filling in the missing bits that will turn *main.c* into our final application.

Here are the main sections:

1. The top portion of the file is spent to define the proper configuration bits according to each processor model needs. For our simple test project the defaults will do.

2. The SPI peripheral initialization follows, this is necessary to enable access to the Serial Flash of the Mikromedia board, the only non volatile memory available, which is used by the Touchscreen routines to store the calibration data (as we have seen in previous chapters).
   Note that this is split into two parts: first an initialization function is defined (in our case this is going to use the SST25 library which is compatible with the M25P80 serial flash); later the function is called and the proper set of parameters is passed to ensure the correct SPI baud rate, clock polarity and active edge mode are selected.

3. The main function is presented. Inside it a call to the `InitializeBoard()` function is performed, followed by the classic GOL main loop scheme.

4. The `GOLMsgCallback()` function is presented. Inside it only a call to the automatically generated actions (`GDDDemoGOLMsgCallback()`).

5. Followed by the `GOLDrawCallback()` function. Inside it once more only a call to automatically generated actions (`GDDDemoGOLDrawCallback()`)

6. The next section is concerned with timing. Timer3 is chosen as the default timer to base all touchscreen sensing events.

7. Finally, the `InitalizeBoard()` function takes up the rest of the *main.c* file. The vast majority of it is actually spent to configure the oscillator (PLL) and I/O pins of the microcontroller. The microcontroller model symbol, predefined in MPLAB X (`__PIC24FJ256GB110__`), is used to select out the major portions but a number of additional symbols are used to help discriminate additional common sections, including `MIKRO_BOARD` in our case.

Out of all these sections, we will need to modify only section 4 and 7:
the `GOLMsgCallback()` and the `InitializeBoard()` function.

Inside the board initialization function, in particular, we will need to configure the OutputCompare module (and PPS) to output a PWM signal onto pin RD2 (the backlight control line), as we did in the previous demo project.

This can be done by calling the same `BacklightInit()` function, appropriately added to the top of *main.c* or just before the `InitializeBoard()` function, as the last action at the very bottom.

```
    ...
    // initialize the components for Resistive Touch Screen
    TouchInit(NVMWrite, NVMRead, NVMSectorErase, TOUCH_INIT_VALUES);

    HardwareButtonInit();            // Initialize the hardware buttons

    // configure OC1 to generate a PWM signal and control the backlight
    BacklightInit();
}   // Initialize Board
```

Listing 7.12 - Modifying the InitializeBoard() function

Also, similarly to what done during the previous demo project, we will insert the desired action inside the `GOLMsgCallback()` body. The exact point of insertion is marked already in the template by a comment line that explicitly enough says: "// Add additional code here ..."

This time though, we want not only the slider value to control the screen luminosity, but we want the Digital Meter widget to be updated too.

Consulting the GOL User Guide, we quickly find the two functions that we need to perform the job:
- `DmSetValue()`, that can be used to update the new numeric value represented by the digital meter widget
- `SetState()`, which we can use to set the DM_UPDATE flag of the Digital Meter widget to force it to update its image on screen

Both functions need a pointer to the Digital Meter object as the first argument, so we can use the `GOLFindObject()` function to identify it based on the known ID (DMT_3).

The updated `GOLMsgCallback()` function can be re-written as follows:

```
WORD GOLMsgCallback(WORD objMsg, OBJ_HEADER *pObj, GOL_MSG *pMsg)
{
    DIGITALMETER* pDM;

    GDDDemoGOLMsgCallback(objMsg, pObj, pMsg);

    // Add additional code here...
    if( pObj->ID == SLD_2 )
    {
        // obtain pointer to Digital meter
        pDM = (DIGITALMETER*) GOLFindObject( DMT_3);

        // read slider position and update the meter
        DmSetValue( pDM, SldGetPos( pObj));

        // force the digital meter to update image on screen
        SetState( pDM, DM_UPDATE);

        // update the screen backlight setting
        BacklightSet( SldGetPos( pObj));
    }
    return (1);
}
```

Listing 7.13 - Customizing the GOLMsgCallback() function

GDD X Checklist

Before launching the project build to test the proper functioning of the new demo code, we need to go through a short checklist to add the folder *Slider* to the *C include dirs* and defining a proper heap size setting.

Open the **Project Properties** dialog box and ensure that:

1. In the *Categories* list, select the **XC16-gcc**

2. In the *Option* categories, select **Preprocessing and Messages**

3. In *C Include Dirs*, enter:

    a) . (dot),

    b) "../Board Support Package"

    c) "../Microchip/Include",

    d) "Slider"

4. In the *Categories* list, select the **XC16-ld**

5. In the *Option* categories, select **General**

6. In *Heap Size*, enter **2048**

At this point we are ready to build the project and verify that all is working as expected. Moving the new vertical slider, verify that both the luminosity of the screen and the value of the Digital Meter are updated accordingly (see Figure 7.8).

Figure 7.8 – GDD Slider screen capture

Summary

The Graphic Object Layer builds on the graphics primitives offered by the Microchip Libraries for Applications by providing a basic set of widgets that can be quickly assembled to form compelling user interfaces for any embedded control application. While it is compatible with most PIC microcontroller models, its performance is better appreciated on 16 and 32-bit models.

We have seen how the PIC24 Mikromedia board is natively supported by the GOL library and how the Graphics Display Designer (GDD X) tool allows for fast prototyping of a graphical user interface and subsequent customization of the actions connected to each visual component.

Tips & Tricks

Editing the GDD templates

During the initial project set up, the GDD tool prepares the basic project structure by copying *GraphicsConfig.h*, *HardwareProfile.h* and *main.c* (and the entire *Configs* folder) from template files. Later during editing, only the files contained in the GDD project folder (*Slider* in our previous example) are updated. Even during the Code Generation cycles, only the files prefixed with GDD_ found in the main project folder are modified automatically. Therefore any edit to the *main.c* module and the two configuration files is guaranteed to remain valid across GDD editing and code generation sessions!
If there are changes to the configuration files (or the main module) that we would like to make available to all future projects, then it is necessary to update the template files themselves. These are found inside the MLA folder structure (they are part of the MLA, not the GDD tool!) in particular inside the folder:
Microchip/Graphics/GDD/Default_Template

Adding Support for a New Board

While most/all of the examples in this book are based on the basic PIC24 Mikromedia board, you might be interested in adding support for newer models and/or customized boards of your own creation. This can be accomplished quite easily by adding a new profile (HWD_*.h file) to the *Configs* template folder, likely after copying and editing one of the many existing examples.

Online Resources

- http://www.microchip.com/gddx

    The home page of the Graphic Display Designer tool

Suggested Reading

- AN1136 – How to Use Widgets in Microchip Graphics Library
- AN1246 – How to Create Widgets in Microchip Graphics Library
- AN1182 – Fonts in the Microchip Graphics Library
- Graphics Display Designer User Guide

Exercises

1. Practice adding *actions* to widgets messages using the built in editor. Notice how the code produced automatically will be inserted in the `GDDDemoGOLMsgCallback()` and `GDDDemoGOLDrawCallback()` defined in the *GDD_X_Event_Handler.c* file.

2. Create a new "Accelerometer Test" project to demonstrate the *ProgressBar* and *RoundDial* widgets and develop a user interface similar to Figure 7.9, to represent the values of the X,Y and Z axis of the Mikromedia on board accelerometer.

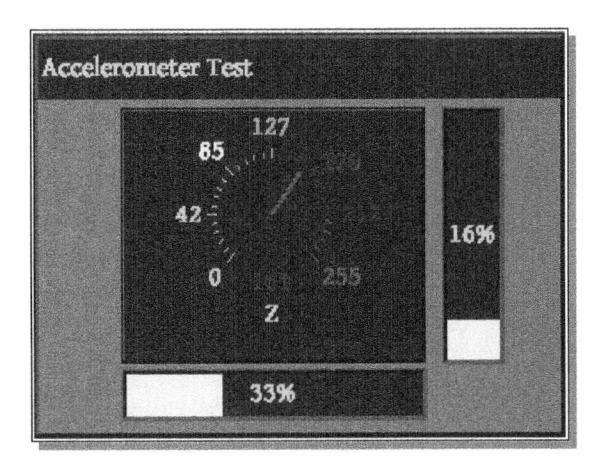

Figure 7.9 – Accelerometer Test screen capture

Solutions

The interface to the on-board Accelerometer, require access to the I2C bus as illustrated in Figure 7.10.

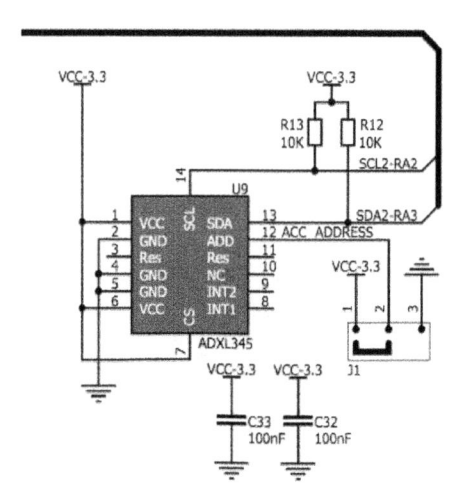

Figure 7.10 – ADXL345 Accelerometer Interface

The following support module (Listing 7.14, 7.15 and 7.16) will allow you to initialize the device and to poll it periodically to retrieve the current values of the X, Y and Z axis acceleration.

```
/*
 * File:    adxl345.h
 */

#ifndef ADXL345_H
#define     ADXL345_H

#define USE_AND_OR

#include <i2c.h>

#define ACC_ADDRESS         0x3A // default address
#define ACC_BAUD_100kHz     157  // Assuming 16MHz clock using Tablle 15-1 DS
#define ACC_DEVID           0x00 // devid register
#define ACC_X0              0x32 // X0 value

BYTE readACCRegister( BYTE reg);
void writeACCRegister( BYTE reg, BYTE b);
int ACCInit( void);
void readACCxyz( int* x, int* y, int* z);


#endif      /* ADXL345_H */
```

Listing 7.14 – ADXL345.h

```c
/*
 * ADXL345.c
 *
 * Accelerometer I2C Interface
 */
#include "ADXL345.h"

void AddressACC( BYTE add)
{
    // 1. write the device addess
    while( 1)
    {
        StartI2C2();    IdleI2C2();
        // send command and address
        MasterWriteI2C2( ACC_ADDRESS);   IdleI2C2();

        // exit if received an acknowledge
        if ( I2C2STATbits.ACKSTAT == 0)
            break;

        StopI2C2();     IdleI2C2();
    } // wait until you receive an acknowledge

    // 2. send the register address
    MasterWriteI2C2( add);     IdleI2C2();
} // address Acc


BYTE readACCRegister( BYTE reg)
{
    BYTE r = 1;

    // 1. select device and register
    AddressACC( reg);

    // 2. issue a read command
    RestartI2C2();    IdleI2C2();
    MasterWriteI2C2( ACC_ADDRESS + 1);   IdleI2C2();

    //3. get one byte of data in
    r = MasterReadI2C2();
    Nop();
    Nop();

    // 4. terminate sequence
    NotAckI2C2(); IdleI2C2();
    Nop();
    Nop();

    StopI2C2(); IdleI2C2();

    // 5. return value read
    return r;

} // read ACC Register


void writeACCRegister( BYTE reg, BYTE b)
{
    AddressACC( reg);
    MasterWriteI2C2( b); IdleI2C2();
    StopI2C2(); IdleI2C2();

} // write ACC register
```

Listing 7.15 – ADXL345.c

```c
/*
 * Accelerometer initialization
 */
int ACCInit( void)
{
    BYTE w;

    // configure I2C port for accelerometer access
    OpenI2C2( I2C_ON | I2C_7BIT_ADD | I2C_STR_EN, ACC_BAUD_100kHz);
    IdleI2C2();

    // 1. test Read DEVID register
    w = readACCRegister( ACC_DEVID);
    if ( w != 0xE5)
        return -1;                       // Failed

    // 2. enable measurement
    writeACCRegister( 0x2D, 0x08 );      // write to POWER_CTL register

    return 0;                            // Success
} // ACC Init

void readACCxyz( int* x, int* y, int* z)
{
    int r;

    // 1. select device and register
    AddressACC( ACC_X0);
    StopI2C2();      IdleI2C2();

    // 2. issue a read command
    StartI2C2();     IdleI2C2();
    MasterWriteI2C2( ACC_ADDRESS + 1); IdleI2C2();

    //3. get X
    r = MasterReadI2C2();            // lsb
    AckI2C2(); IdleI2C2();           // ask for more
    r |= ( MasterReadI2C2() << 8);   // msb
    *x = r;
    AckI2C2(); IdleI2C2();           // ask for more

    //4. get Y
    r = MasterReadI2C2();            // lsb
    AckI2C2(); IdleI2C2();           // ask for more
    r |= ( MasterReadI2C2() << 8);   // msb
    *y = r;
    AckI2C2(); IdleI2C2();           // ask for more

    //5. get Z
    r = MasterReadI2C2();            // lsb
    AckI2C2(); IdleI2C2();           // ask for more
    r |= ( MasterReadI2C2() << 8);   // msb
    *z = r;

    // 6. terminate sequence
    NotAckI2C2(); IdleI2C2();
    StopI2C2();   IdleI2C2();

} // readACCxyz
```

Listing 7.16 – ADXL345.c (continued)

Chapter 8

USB

The applications for USB connectivity in embedded control are so numerous and varied that it is impossible to give justice to the matter in a single chapter. In fact, an entire book would not suffice. Many books have been written, by more qualified authors, to help you understand the inner workings, the physical layer and the protocols used to implement various classes of USB applications. So I will assume that you will be able to find any additional information required to satisfy your curiosity in the suggested readings at the end of this chapter. I will instead spend the few pages available here, to focus on the *use* of USB, and even that will be limited to a very specific subset of all possible applications: the connection of an embedded device to a PC as a replacement for the *good old serial port*.

A Very Brief Introduction to USB

Let's summarize, in as few paragraphs as possible, what is the absolute minimum that you need to know about USB to be effective. Let's also limit the conversation to the sole Low Speed and Full Speed USB standards as they are the only ones applicable to the PIC24 (and PIC32) microcontroller family as of this writing.
First of all there is an absolute master, the *host*, which is going to be a PC in most cases, and every other node on the bus is a slave or *device* to use USB proper terminology. Only the host can initiate a conversation. It gets to decide who can send data back to him, how much and when. The entire protocol design is meant to simplify operations for the host (the PC), not the (embedded) device, so to best handle the workload.
As a result, most embedded control designers, confronting USB connectivity applications for the first time, are surprised by the complexity of the system compared to the freedom and simplicity afforded by the serial port they might be migrating from.

The good news is that the MLA comes to our rescue with a large library (contained in the /USB folder) that covers most of the applications classes we might need.
The MLA USB support also includes the possibility to use the embedded device in the role of the host, a very useful feature when there is no PC involved in the application and the microcontroller has to take control of one or more devices such as memory sticks, joysticks and printers.

Coding with Class

In an effort of simplification, USB applications are grouped in *classes* with strictly defined conventions on how data is supposed to be packaged and transmitted. There is room for up to 256 possible classes of applications in the USB standard, but Table 8.1 summarizes the few that we are more likely to encounter in the wild.

Class Name	Applications Examples	Default Drivers Available Win	OS X	Linux	MLA Support
Audio	Speaker, microphone, sound card, MIDI	✓	✓	✓	✓
Communications Device (CDC)	Modem, Ethernet adapter, Wi-Fi adapter	✓ ([4])	✓	✓	✓
Human interface device (HID)	Keyboard, mouse, joystick	✓	✓	✓	✓
Physical Interface Device (PID)	Force feedback joystick				✓
Image	Camera, scanner				
Printer	Laser printer, inkjet printer, CNC machine	✓	✓	✓	✓
Mass storage (MSC or UMS)	USB flash drive, memory card reader, digital audio player, digital camera, external drive	✓	✓	✓	✓
USB hub	Full bandwidth hub	✓	✓	✓	✓
Smart Card	USB smart card reader	✓	✓	✓	✓
Content security	Fingerprint reader				
Video	Webcam				
Personal Healthcare	Pulse monitor (watch)				✓
Audio/Video (AV)	Webcam, TV				
Wireless Controller	Bluetooth adapter, Microsoft RNDIS				
Vendor-specific	Indicates that a device needs vendor specific drivers				✓

Table 8.1 – USB classes

When an application fits perfectly in the definition of one of the above classes, for example a Human Interface Device, a.k.a. a mouse, we are in luck. The standard defines very clearly how data is supposed to be packaged, and all the special needs of the application are already taken into consideration and supported ideally. Unfortunately most of the "interesting" embedded control projects are not so easily classified and therefore require some more research and often a bit of creativity.

---

[4]) Windows users will find that even if the CDC class driver is already available and installed on their machine, upon first connection of a CDC device, they will be asked to provide a ".inf" file. This is NOT a driver, but just a text file that contains additional information that the operating system will use to better identify the new (virtual) serial port. A template file, *mchpcdc.inf,* can be found in the */USB/Device-CDC-Basic Demo/inf* folder of the MLA.

The Drivers Problem

Finding the right class for an embedded application is particularly important for one reason, the one thing that keeps most embedded control designers awake at night: the need for *drivers*!

Each of the application classes defined by the USB standard requires that a matching driver be loaded by the (PC) operating system to handle the device as soon as it detects its presence on the bus. This is a particularly delicate piece of code that needs to operate at the OS lowest level (kernel).

Common operating systems (Windows, OS X, and Linux) include a good number of such class drivers ready to be loaded, but not for all classes and not in a consistent way across platforms (see Table 8.1 Default drivers column). This means that in many cases the embedded control designer is faced with the need to develop custom (possibly multiple) operating system drivers to support his application.

Writing an application for a popular PC operating system does require a different skill set (and tools set) from the ones possessed by the average embedded control designer (i.e. MPLAB® X and the XC16 compiler won't suffice). But writing a "driver" for Windows, OS X or Linux is a challenge of an order of magnitude greater and requires a much more rare skill set.

Unless you are one such skilled programmer, there are only three ways to solve this problem:

1. Trying to make the application "fit" into one of the classes that have driver support built into the target operating system(s)
2. Using the MLA Generic Device Drivers examples and trying to customize it (this will cover only the Windows platform)
3. Relying on third parties to develop the drivers and/or use third party driver's design kits (expensive)

Aiming at the simplest and most common use cases, in the rest of this chapter we will explore only the first of the three options!

The Physical Layer

USB devices have become so ubiquitous that I am sure we don't need to spend many words to describe the exterior physical aspect of USB connectors and cables, as we are all constantly surrounded by a multitude of chargers and adapters, we get to know them all too well. The reader might be less familiar though with the inner detail.

A typical USB cable will include four copper wires (see Table 8.2). Two are dedicated to provide power to the device while the remaining two form a differential pair (D+ and D-). The pair is usually twisted and then further shielded and presents a characteristic impedance of 90 Ohm.

While the supply provided by the two power lines is 5V (+/- 5%), the differential pair operates with 3V logic.

Pin	Name	Cable color	Description
1	VBUS	Red (or Orange)	+5V
2	D-	White (or Gold)	Data -
3	D+	Green	Data +
4	GND	Black (or Blue)	Ground

Table 8.2 – USB cables detail

USB data is transmitted by toggling the data lines between the J state (D- low and D+ high) and the opposite K state (D- high and D+ low). USB encodes data using the NRZI line coding; a 0 bit is transmitted by toggling the data lines from J to K or vice-versa, while a 1 bit is transmitted by leaving the data lines unchanged.

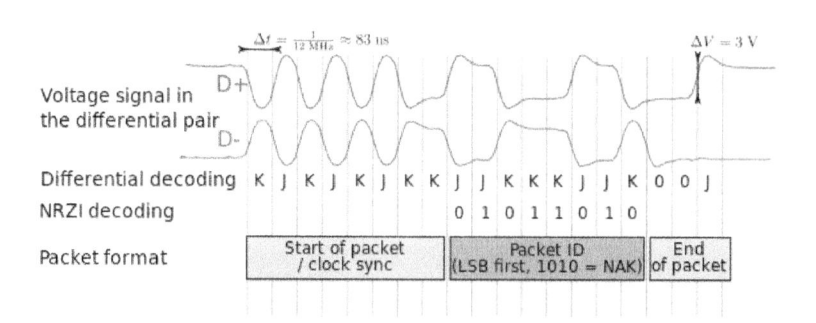

Figure 8.1 – USB Signaling Example

To ensure a minimum density of signal transitions remains in the bitstream, USB uses *bit stuffing*; an extra 0 bit is inserted into the data stream after any appearance of six consecutive 1 bits. This way a sequence of seven consecutive received 1 bits (differential pair unchanged for seven clock periods or more) is always a sign of error.

Special signaling sequences, called *tokens*, are used to separate different phases of the communication. Note that tokens are not restricted to using exclusively the J and K states.

Use of the bus is controlled by the host and split in *frames* of finite duration. The host starts a new frame by sending a special *Start of Frame* token every 1ms for Full Speed connections and every 10ms for Low Speed connections. Devices are allowed to respond, providing handshake and eventually data back to the host, only when specifically addressed by the host and within strictly defined timing boundaries.

Power
As the device is connected to the bus, it gets immediately access to the 5V power supply up to a guaranteed maximum of 100mA. It is only after the logical connection has been established with the host though that it is possible to require more current, up to 5 times as much (500mA).

Note that the request for more power can be refused by the host depending on the power budget available, current loading of the PC or by specific limitation of the nodes (hubs) eventually interposed. Further, when necessary, the host can require the device to enter a so called *suspended* mode, reducing its power consumption to the lowest possible amount compatible with the application. Later, it will signal the device when to resume normal operation, providing a minimum amount of time to comply (10ms).

USB devices can also be *self-powered*, in which case they must ensure not to supply current back onto the bus power lines.

Beginning from the End(point)
All the hardware/physical details we have quickly reviewed in the previous section disappear from our view when we look at the USB bus from a software perspective.
A typical microcontroller USB peripheral contains a Serial Interface Engine (SIE), which takes a role similar to that of the UART in asynchronous serial communication and is responsible for the translation of the physical signals on the bus wires into data moving back and forth.

Data moves from the host to a device (*out*) and vice versa (*in*) in packets whose size is fixed and determined during an initial phase of handshake (*enumeration*), which we will explore in more details shortly. In fact, contrary to what happens in a UART, it's packets of data, not single bytes, that move from point to point, PC to device or device to PC, in discrete (block) transactions.

The term *endpoint* is used in the USB standard to define the containers of such packets, buffers in RAM. Endpoints are mono-directional. They can either be used for *in* transactions (from device to PC) or *out* transactions (PC to device).

NOTE
> Notice how the terminology *in/out* used in USB literature is always relative to the *host*, the benign dictator ruling the bus.

Access to the content of an endpoint must be carefully controlled to avoid conflicts between the microcontroller and the USB peripheral (SIE). Flags are used by the SIE to signal when an endpoint is in "use", meaning its content is being read or written to by the USB peripheral, or it is available, so that the microcontroller can access its contents.

In order to transfer a block of data to the PC, a typical device will fill an (*in*) endpoint buffer with up to the allowed number of bytes and then will communicate to the SIE that the endpoint is *armed* and ready to send data. No more data can be sent to the PC until the SIE has emptied the endpoint buffer and released it.

Similarly, when expecting data from the host, a microcontroller will poll the status of an (*out*) endpoint until the SIE has filled it with a new batch of data and released it.

Each device on the USB bus can establish multiple simultaneous bidirectional connections with the host by defining up to 32 endpoints (0-15 *in* and 0-15 *out* endpoints).

The SIE will handle the data to and from all the endpoints defined by multiplexing it in the data stream with the host creating effectively up to 16 virtual connections between host and device.

It will be the host though to decide which endpoints to serve, and how often, based on the available bandwidth, and the processor (and OS) workload, while attempting to satisfy the device "needs".

Types of Transfers

Each endpoint can be assigned one of four possible transfer types:
- *Control*, used for identification and configuration
- *Interrupt*, for small blocks of data that must be delivered timely
- *Bulk*, used by storage devices for example, for large blocks of data whose integrity must be guaranteed but whose latency is not critical
- *Isochronous*, used by audio streaming devices for example, for large blocks of data with guaranteed delivery latency but whose integrity need not be guaranteed

A host will intersperse transfers of all four types in each frame, and will use the endpoint transfer type to decide the order and relative priority according to a detailed set of rules defined in the USB standard.

Enumeration

In order for the device and host to establish the correct links between the endpoints that will compose the connection, it is crucial that the two have a good "talk" as soon as they "meet" each other on the bus. This phase is called the *enumeration*. It is performed each time a host start taking control of a bus, and repeated each time a new device is attached. During this phase the host queries the new device(s) individually, assigns it an *address* and then invites the device to make "requests".

The enumeration uses always the two (*control* type) endpoints 0-*in* and 0-*out* and starts with the device being enumerated sending a small list of tables, known as *descriptors,* packed sequentially in control endpoint *0-in*. The rules that define the interpretation of the descriptors tables content are specified in extreme detail by the USB standard. During the enumeration dialog, a device can provide important information about its manufacturer, model, serial number and data transfer timing requirements. Endpoints are grouped logically in *interfaces* according to their function and interfaces are grouped in *configurations*. Multiple configurations options can be offered to the host so to allow the maximum flexibility to adapt to different operating system and host capabilities.

Eventually all this information is used by the host to choose the optimal drivers to use with the device. It is here that the USB application classes we introduced at the beginning of this chapter come in handy. They allow us to simplify somewhat the process and reduce it to more manageable terms by defining common rules for classes of similar applications and allowing the host to share/re-use device drivers whenever possible.

We don't have the space to go into further detail here, but it is important to close this brief introduction noting that as part of the MLA distribution, tools are included to allow us to create descriptors tables with ease and according to the standards. In most cases, we will be able to simply re-use standard descriptors template/demo files offered in the many demonstration projects.

MLA USB support modules

All the necessary symbols definitions and function prototypes are conveniently collected for us by one master header file: ***usb.h*** found in the */USB* subfolder of the **Include** path.

The enumeration process is typically managed by one complex state machine as it is composed of multiple transactions in both directions between host and device. Fortunately this is a complex piece of code that we don't have to write and most importantly debug. The MLA USB library provides all the required logic, in a processor independent form, in one single library module: ***usb_device.c*** found in the */USB* source subfolder.

The *usb_device.c* module contains already all the logic required by a device to handle regular transactions as well, so this module constitutes the true foundation of all our future applications.

There is then an additional module that will be required in order to support specifically the PIC24 family of microcontrollers: ***usb_hal_pic24.c*** (the acronym *_hal* stands for hardware abstraction layer).

In our case, there is only one function that is provided by the PIC24 HAL, and it is covering a special case when a device is requested to enter a low power mode by the host (Sleep on Suspend).

Since our primary goal was defined as the pure replacement of the "good old serial port", we can focus our efforts on two specific USB classes:

- the *Communication Device* class or *CDC*
- the *Human Interface Device* class or *HID*

The first one, CDC, might seem the most obvious choice. After all it was meant to interface to modems and other (serial) communication interfaces. So let's start exploring a simple CDC application to evaluate its pros and cons.

We will take a look at the HID class later on in the second part of this chapter.

Communication Device Class

By choosing to conform our application to the CDC class, a number of configuration choices are made for us. Granted the full specifications of the CDC class can accommodate a large number of communication devices that include telephone modems, ISDN modems, Ethernet interfaces and Wireless interfaces, once we restrict it to the specific case of the interface to a very basic modem, we get a pretty simple recipe (*interface*):

- Endpoint 0: (*control*) in/out is (always) used to enumerate and set up the connection
- Endpoint 1: (*interrupt*) in, is used to provide basic modem status information
- Endpoint 2: (*bulk*) in/out, is used to move packets of actual data to and from the device

It is the latter pair of endpoints that defines an important characteristic of a CDC class interface. Since data is transferred using *bulk* transactions, the integrity of the data is guaranteed (host and device will verify it using CRC codes on each packet) but the latency will not be guaranteed.

This means that although we will get a high (average) bandwidth, the actual time of arrival of each single packet of data will not be deterministic. Some packets might be delayed more than others depending on the bus loading conditions, the number and size of simultaneous data transfers requested by our or other devices on the bus.

Another benefit of using the CDC class is that we can use immediately as a template, the file ***usb_descriptors.c*** in Listings 8.1, containing the descriptors tables as defined in the MLA demonstration project **CDC-Device.**

CDC Descriptors

The *usb_descriptors.c* file is normally prepared using an automated tool but it is worth inspecting its contents to gain some insight in the function and composition of the *descriptors tables* that characterize a typical CDC class application.

> **NOTE**
>
> The *usb_descriptors.c* file is not a header file (.h) but a source (.c) file since the data structures that it defines are also initialized.
>
> We need to add the file to the project in the *Source Files* logical folder!

```
1.      ifndef __USB_DESCRIPTORS_C
2.      #define __USB_DESCRIPTORS_C
3.
4.      #include "./USB/usb.h"
5.      #include "./USB/usb_function_cdc.h"
6.
7.      /* Device Descriptor */
8.      ROM USB_DEVICE_DESCRIPTOR device_dsc=
9.      {
10.         0x12,                       // Size of this descriptor in bytes
11.         USB_DESCRIPTOR_DEVICE,      // DEVICE descriptor type
12.         0x0200,                     // USB Spec Release Number in BCD format
13.         CDC_DEVICE,                 // Class Code
14.         0x00,                       // Subclass code
15.         0x00,                       // Protocol code
16.         USB_EP0_BUFF_SIZE,          // Max packet size for EP0, see usb_config.h
17.         0x04D8,                     // Vendor ID
18.         0x000A,                     // Product ID: CDC RS-232 Emulation Demo
19.         0x0100,                     // Device release number in BCD format
20.         0x01,                       // Manufacturer string index
21.         0x02,                       // Product string index
22.         0x00,                       // Device serial number string index
23.         0x01                        // Number of possible configurations
24.     };
25.
26.     /* Configuration 1 Descriptor */
27.     ROM BYTE configDescriptor1[]={
28.         /* Configuration Descriptor */
29.         0x09,//sizeof(USB_CFG_DSC),    // Size of this descriptor in bytes
30.         USB_DESCRIPTOR_CONFIGURATION,  // CONFIGURATION descriptor type
31.         67,0,                          // Total length of data for this cfg
32.         2,                             // Number of interfaces in this cfg
33.         1,                             // Index value of this configuration
34.         0,                             // Configuration string index
35.         _DEFAULT | _SELF,              // Attributes, see usb_device.h
36.         50,                            // Max power consumption (2X mA)
37.
```

Listing 8.1a - *usb_descriptors.c*, Device Descriptor table

The *device descriptor* table specifies several important elements, among which:

- line 13, the class to be used
- line 17, the Vendor ID, in this case Microchip Technology own
- line 18, the ProductID, a code used for the RS232 Emulation Demo

NOTE

> We will discuss how to obtain our own Vendor ID and how to choose a Product ID code in the Tips and Tricks section at the end of this chapter.

- lines 20, 21, 22, strings defining the manufacturer serial number and product name
- line 23, tells the host that there is only one configuration possible

The *configuration descriptor* table adds several more items:

- line 26, the configuration is composed of two interfaces

- line 36, the maximum current requested is going to be 100mA

    A larger value can be requested, up to 500mA, by modifying this field as needed.

```
38.         /* Interface Descriptor */
39.         9,//sizeof(USB_INTF_DSC),   // Size of this descriptor in bytes
40.         USB_DESCRIPTOR_INTERFACE,   // INTERFACE descriptor type
41.         0,                          // Interface Number
42.         0,                          // Alternate Setting Number
43.         1,                          // Number of endpoints in this intf
44.         COMM_INTF,                  // Class code
45.         ABSTRACT_CONTROL_MODEL,     // Subclass code
46.         V25TER,                     // Protocol code
47.         0,                          // Interface string index
```

Listing 8.1b – *usb descriptors.c* , First Interface Descriptor table

The first *interface descriptor table* adds:

- line 43, one endpoint is used

- line 44, 45, 46 to support a communication interface with particular control signals here referred to as the "V25 terminal" protocol.

```
48.
49.         /* CDC Class-Specific Descriptors */
50.         sizeof(USB_CDC_HEADER_FN_DSC),
51.         CS_INTERFACE,
52.         DSC_FN_HEADER,
53.         0x10,0x01,
54.
55.         sizeof(USB_CDC_ACM_FN_DSC),
56.         CS_INTERFACE,
57.         DSC_FN_ACM,
58.         USB_CDC_ACM_FN_DSC_VAL,
59.
60.         sizeof(USB_CDC_UNION_FN_DSC),
61.         CS_INTERFACE,
62.         DSC_FN_UNION,
63.         CDC_COMM_INTF_ID,
64.         CDC_DATA_INTF_ID,
65.
66.         sizeof(USB_CDC_CALL_MGT_FN_DSC),
67.         CS_INTERFACE,
68.         DSC_FN_CALL_MGT,
69.         0x00,
70.         CDC_DATA_INTF_ID,
71.
```

Listing 8.1c – *usb descriptors.c* , CDC Class Specific Descriptors table

The *CDC class specific descriptors* table is much harder to decipher, but we can see in it (lines 63 and 64) that it makes reference to two communication interfaces, a "CDC_COMM" and a "CDC_DATA" interface.

```
72.         /* Endpoint Descriptor */
73.         //sizeof(USB_EP_DSC),DSC_EP,_EP02_IN,_INT,CDC_INT_EP_SIZE,0x02,
74.         0x07,/*sizeof(USB_EP_DSC)*/
75.         USB_DESCRIPTOR_ENDPOINT,    //Endpoint Descriptor
76.         _EP01_IN,                    //EndpointAddress
77.         _INTERRUPT,                  //Attributes
78.         0x08,0x00,                   //size
79.         0x02,                        //Interval
80.
```

Listing 8.1d – *usb descriptors.c* , First Endpoint Descriptor table

In Listing 8.1d the first endpoint is defined for the CDC_COM interface.
It uses endpoint *1-in* (line 76) and its type is *interrupt* (line 77).
It can transfer 8 bytes of data at a time (line 78).

```
81.         /* Interface Descriptor */
82.         9,//sizeof(USB_INTF_DSC),   // Size of this descriptor in bytes
83.         USB_DESCRIPTOR_INTERFACE,            // INTERFACE descriptor type
84.         1,                          // Interface Number
85.         0,                          // Alternate Setting Number
86.         2,                          // Number of endpoints in this intf
87.         DATA_INTF,                  // Class code
88.         0,                          // Subclass code
89.         NO_PROTOCOL,                // Protocol code
90.         0,                          // Interface string index
91.
92.         /* Endpoint Descriptor */
93.         //sizeof(USB_EP_DSC),DSC_EP,_EP03_OUT,_BULK,CDC_BULK_OUT_EP_SIZE,0x00,
94.         0x07,/*sizeof(USB_EP_DSC)*/
95.         USB_DESCRIPTOR_ENDPOINT,    //Endpoint Descriptor
96.         _EP02_OUT,                   //EndpointAddress
97.         _BULK,                       //Attributes
98.         0x40,0x00,                   //size
99.         0x00,                        //Interval
100.
101.        /* Endpoint Descriptor */
102.        //sizeof(USB_EP_DSC),DSC_EP,_EP03_IN,_BULK,CDC_BULK_IN_EP_SIZE,0x00
103.        0x07,/*sizeof(USB_EP_DSC)*/
104.        USB_DESCRIPTOR_ENDPOINT,    //Endpoint Descriptor
105.        _EP02_IN,                    //EndpointAddress
106.        _BULK,                       //Attributes
107.        0x40,0x00,                   //size
108.        0x00,                        //Interval
109.    };
```

Listing 8.1e – *usb descriptors.c* , First Endpoint Descriptor table

The second and last *interface descriptors* table (Listing 8.1e) tells us that:

- line 86, it requires two endpoints
- lines 97 and 106, they are both of type *bulk*

- lines 98 and 107, they can transfer up to 64 bytes (0x40) each, the maximum allowed for a bulk endpoint

From this last item we deduce what is going to be the theoretical maximum data transfer rate achievable using this configuration.

Since we won't be able to send or receive more than one bulk endpoint per USB frame (1 frame = 1ms in Full Speed mode) it follows that:

$$\text{Bitrate}_{MAX} = 64 * 8 * 1000 = 512{,}000 \text{ bit/second}$$

Although this is an absolute maximum value, we can see how our data transfer capability places us comfortably (four times) above that of a typical serial port which would max out at around 115.2 kBaud.

```
110.    //Language code string descriptor
111.    ROM struct{BYTE bLength;BYTE bDscType;WORD string[1];}sd000={
112.    sizeof(sd000),USB_DESCRIPTOR_STRING,{0x0409}};
113.
114.    //Manufacturer string descriptor
115.    ROM struct{BYTE bLength;BYTE bDscType;WORD string[25];}sd001={
116.    sizeof(sd001),USB_DESCRIPTOR_STRING,
117.    {'M','i','c','r','o','c','h','i','p',' ',
118.    'T','e','c','h','n','o','l','o','g','y',' ','I','n','c','.'
119.    }};
120.
121.    //Product string descriptor
122.    ROM struct{BYTE bLength;BYTE bDscType;WORD string[25];}sd002={
123.    sizeof(sd002),USB_DESCRIPTOR_STRING,
124.    {'C','D','C',' ',' ','R','S','-','2','3','2',' ',' ',
125.    'E','m','u','l','a','t','i','o','n',' ',' ','D','e','m','o'}
126.    };
127.
128.    //Array of configuration descriptors
129.    ROM BYTE *ROM USB_CD_Ptr[]=
130.    {
131.        (ROM BYTE *ROM)&configDescriptor1
132.    };
133.    //Array of string descriptors
134.    ROM BYTE *ROM USB_SD_Ptr[USB_NUM_STRING_DESCRIPTORS]=
135.    {
136.        (ROM BYTE *ROM)&sd000,
137.        (ROM BYTE *ROM)&sd001,
138.        (ROM BYTE *ROM)&sd002
139.    };
140.    #endif
```

Listing 8.1f – *usb_descriptors.c* **, String tables**

The *usb_descriptors.c* file ends with a few more tables that contain the manufacturer name and product description as UNICODE strings.

NOTE

> UNICODE strings are defined as arrays of WORD rather than char type. Therefore each character must be spelled individually, surrounded by single quotes, and each string length must be declared explicitly.

Using the CDC class

The main CDC class support module: ***usb_function_cdc.c*** can be found in the **Microchip/USB/CDCDeviceDriver** folder. This module adds to the *usb_device.c* and *usb_hal_pic24.c* modules all the functions and features that are unique to a CDC class application. It must be added, as the previous two, to the Source Files logical folder of a project (likely inside a USB subfolder).

As in all other MLA libraries, we will be expected to call an initialization function:

```
USBDeviceInit();        // Initializes USB module SFRs and firmware
```

before entering the main application loop.

Once in the main loop of the application, there will be two possible modes of operation: *polling* and *interrupt*.

In *polling mode*, all the monitoring activity of the USB engine must be performed by repeatedly calling the function USBDeviceTasks() in the main loop. This imposes some pretty significant limitations in the design of the entire application. In other words, the main loop must be guaranteed to execute fast enough to allow the USB engine state machine to keep up with the protocol demands, which means that all other application tasks must be carefully *fragmented* so that the sum of all the tasks executed in each loop is never exceeding the duration of a USB frame. During normal operation this means 1 ms assuming Full Speed operation, or 10 ms in Low Speed mode.

Polling mode is probably the most efficient way to design an 8-bit application, particularly in a small device with limited interrupt capabilities. But for 16 and 32-bit microcontrollers with a rich set of interrupts, shadow registers for fast context switch and advanced priority control mechanisms, we will rather choose interrupt mode.

In *interrupt mode*, the USB library operates more "automatically" performing all the state machine monitoring and updates in the background using the USB own interrupt vector. The corresponding interrupt service routine is already provided by the USB device library module (*usb_device.c*) and it is activated whenever a USB event is triggered.

As we have seen in all previous MLA libraries, a configuration (header) file is typically used to select among available options. So it is no surprise that the choice between the two modes can be performed via a #define switch in the configuration header file: ***usb_config.h***

Inspecting usb_config.h

In all truth, the *usb_config.h* file is more than just a collection of #define switches. It does cover also the important role of linking some of the data defined in the *usb_descriptors.c* file to the rest of the USB device library.

Listing 8.2 contains a shortened version of *usb_config.h* we will use in our application, where most of the verbose comments have been removed. As for the descriptors file, we do not need to create a new file from scratch each time, we can use the usb_config header files from the MLA example projects as a template and modify it to our liking.

```
1.      /********************************************************************
2.       FileName:         usb_config.h
3.       Dependencies:     GenericTypeDefs.h, usb_device.h
4.
5.      /********************************************************************
6.       * Descriptor specific type definitions are defined in: usbd.h
7.       ********************************************************************/
8.
9.      #ifndef USBCFG_H
10.     #define USBCFG_H
11.
12.     /** DEFINITIONS ****************************************************/
13.     #define USB_EP0_BUFF_SIZE   8     // Valid Options: 8, 16, 32, or 64 bytes.
14.
15.     #define USB_MAX_NUM_INT     2     // For tracking Alternate Setting
16.
17.     //Device descriptor - if these two definitions are not defined then
18.     //   a ROM USB_DEVICE_DESCRIPTOR variable by the exact name of device_dsc
19.     //   must exist.
20.     #define USB_USER_DEVICE_DESCRIPTOR &device_dsc
21.     #define USB_USER_DEVICE_DESCRIPTOR_INCLUDE extern ROM USB_DEVICE_DESCRIPTOR device_dsc
22.
23.     //Configuration descriptors - if these two definitions do not exist then
24.     //   a ROM BYTE *ROM variable named exactly USB_CD_Ptr[] must exist.
25.     #define USB_USER_CONFIG_DESCRIPTOR USB_CD_Ptr
26.     #define USB_USER_CONFIG_DESCRIPTOR_INCLUDE extern ROM BYTE *ROM USB_CD_Ptr[]
27.
28.     //#define USB_PING_PONG_MODE USB_PING_PONG__NO_PING_PONG
29.     #define USB_PING_PONG_MODE USB_PING_PONG__FULL_PING_PONG
30.     //#define USB_PING_PONG_MODE USB_PING_PONG__EP0_OUT_ONLY
31.     //#define USB_PING_PONG_MODE USB_PING_PONG__ALL_BUT_EP0
32.
33.     //#define USB_POLLING
34.     #define USB_INTERRUPT
35.
36.     /* Parameter definitions are defined in usb_device.h */
37.     #define USB_PULLUP_OPTION USB_PULLUP_ENABLE
38.     //#define USB_PULLUP_OPTION USB_PULLUP_DISABLED
39.
40.     #define USB_TRANSCEIVER_OPTION USB_INTERNAL_TRANSCEIVER
41.     //#define USB_TRANSCEIVER_OPTION USB_EXTERNAL_TRANSCEIVER
42.
43.     #define USB_SPEED_OPTION USB_FULL_SPEED
44.     //#define USB_SPEED_OPTION USB_LOW_SPEED
45.
46.     //-------------------------------------------------------------------
47.     #define USB_ENABLE_STATUS_STAGE_TIMEOUTS
48.     #define USB_STATUS_STAGE_TIMEOUT      (BYTE)45    //Approximate timeout in ms
```

Listing 8.2a - usb_config.h

This segment of the *usb_config.h* file contains most of the general configuration switches:

- lines 20-26, specify that we intend to place the descriptor tables in the PIC24 flash memory. Some applications might want to use pointers in RAM instead so that the device can change dynamically the contents of the tables and enumerate in different ways depending on the application status.

- Lines 28-32, select the ping pong mode for all endpoints. This feature is available on the PIC24 USB engine and allows for a better utilization of the buffers, resulting in faster data transfers.

- Lines 33 and 34, perform the selection between interrupt and polling mode

- Line 37, enables the use of the PIC24 internal USB pull up resistor on the D+ pin. This is used to indicate to the host the desired operating speed, when first attached to the bus avoiding the need for any external components

- Line 40, selects the use of the internal USB Full Speed Transceiver, optionally an external one could be selected if desired.

- Line 43, selects the USB SIE desired operating speed, Full Speed in our case.

```
49.     //----------------------------------------------------------------------------
50.     #define USB_SUPPORT_DEVICE
51.
52.     #define USB_NUM_STRING_DESCRIPTORS 3
53.
54.     #define USB_ENABLE_ALL_HANDLERS
55.
56.     /** DEVICE CLASS USAGE **************************************/
57.     #define USB_USE_CDC
58.
59.     /** ENDPOINTS ALLOCATION ************************************/
60.     #define USB_MAX_EP_NUMBER          2
61.
62.     /* CDC */
63.     #define CDC_COMM_INTF_ID       0x00
64.     #define CDC_COMM_EP               1
65.     #define CDC_COMM_IN_EP_SIZE      10
66.
67.     #define CDC_DATA_INTF_ID       0x01
68.     #define CDC_DATA_EP               2
69.     #define CDC_DATA_OUT_EP_SIZE     64
70.     #define CDC_DATA_IN_EP_SIZE      64
71.
72.     //#define USB_CDC_SUPPORT_ABSTRACT_CONTROL_MANAGEMENT_CAPABILITIES_D2
        //Send_Break command
73.     #define USB_CDC_SUPPORT_ABSTRACT_CONTROL_MANAGEMENT_CAPABILITIES_D1
        //Set_Line_Coding, Set_Control_Line_State, Get_Line_Coding, and Serial_State
        commands
74.
75.     #endif //USBCFG_H
```

Listing 8.2b – usb_config.h (continued)

This second section of the usb_config.h header file is mostly related to CDC specific options:

- Line 50, we are going to implement a USB device (as opposed to a host)
- Line 54, allows us to define hooks for all USB event handlers (more on this shortly)
- Line 57, selects the CDC class
- Lines 62-74, define symbols that will be used in the CDC driver and application code

NOTE

> Ensure that these definitions ALWAYS match the contents of the *usb_descriptors.c* file. Any discrepancy here will cause the host and the device to operate under different assumptions with rather unpredictable results.

Getting Attached

Once the USB peripheral is initialized, it is time to check if a connection with the host can be established. If an application is taking its power directly from the USB bus this is granted, but if the application is self-powered this can be revealed only by checking first the status of the VBUS signal. In both cases we can use the function: `USBDeviceAttach()` to initiate the operation. It will ensure that a proper pull up resistor is activated, alerting the host of the device *presence* on the bus and triggering the enumeration process.
The enumeration process will proceed automatically from here thanks to a state machine implemented inside the *usb_device.c* module.

We can keep track of its progress by inspecting the current state thanks to the function: `USBGetDeviceState()`. This is going to proceed through the following discrete states:

- `DETACHED_STATE`, before a connection is attempted
- `ATTACHED_STATE`, before the port/hub it is connected to is configured
- `POWERED_STATE`, after the port/hub is configured and provides power to the device
- `DEFAULT_STATE`, expecting to receive an address from the host
- `ADDRESS_STATE`, after receiving an address
- `CONFIGURED_STATE`, fully configured and ready to communicate

Only the last state represents a condition where true and complete connection has been established with the host and data can begin to flow between the two sides. In fact in most applications it won't make sense to proceed beyond this point until the configured state has been reached.

If we use the LCD Terminal module, developed in the previous chapters, we could make this sequence "visible" to the user by logging on the Mikromedia board screen each state transition by calling in a loop the DisplayUSBStatus() function shown in Listing 8.3.

```
void DisplayUSBStatus(void)
{
    static unsigned state = 0x1234;

    // check if status changed
    if (USBDeviceState == state)  return;

    state = USBDeviceState;

    // check if application connected but suspended
    if( USBSuspendControl == 0)
    {
        LCDClear();

        if(USBDeviceState == DETACHED_STATE)
        {
            LCDPutString( "Detached\n");
        }
        else if(USBDeviceState == ATTACHED_STATE)
        {
            LCDPutString( "Attached\n");
        }
        else if(USBDeviceState == POWERED_STATE)
        {
            LCDPutString( "Powered\n");
        }
        else if(USBDeviceState == DEFAULT_STATE)
        {
            LCDPutString( "Default\n");
        }
        else if(USBDeviceState == ADDRESS_STATE)
        {
            LCDPutString( "Address\n");
        }
        else if(USBDeviceState == CONFIGURED_STATE)
        {
            LCDPutString( "Connected\n");
        }
    }
}// DisplayUSBStatus
```

Listing 8.3 – DisplayUSBStatus() function

Virtual Serial Ports

Once the connection is established, if the host is running the Windows operating system, this will result in the creation of a *virtual serial port* that will appear numbered sequentially among the other COMx ports.

If the host is running OS X or Linux, similar *virtual serial ports* will appear in the */dev* folder as *tty-xxx* devices or *con-xxx* devices.

From the host operating system and its applications there is absolutely no difference between a "good old serial port" and the new "virtual serial port" connection just established.

The same applications (Hyperminal, Teraterm, CoolTerm...) and utilities that are used to access the standard (legacy) serial ports, will operate transparently with the virtual serial ports once a Mikromedia board is connected to the USB bus using the CDC class.

This means no special coding/hacking is required to interface to legacy PC applications allowing for maximum re-use of knowledge and existing tools on the host side.

This is the single greatest advantage deriving from the use of the CDC class!

Figure 8.2 shows how the Mikromedia board appears to CoolTerm, a popular terminal emulation application for OS X.

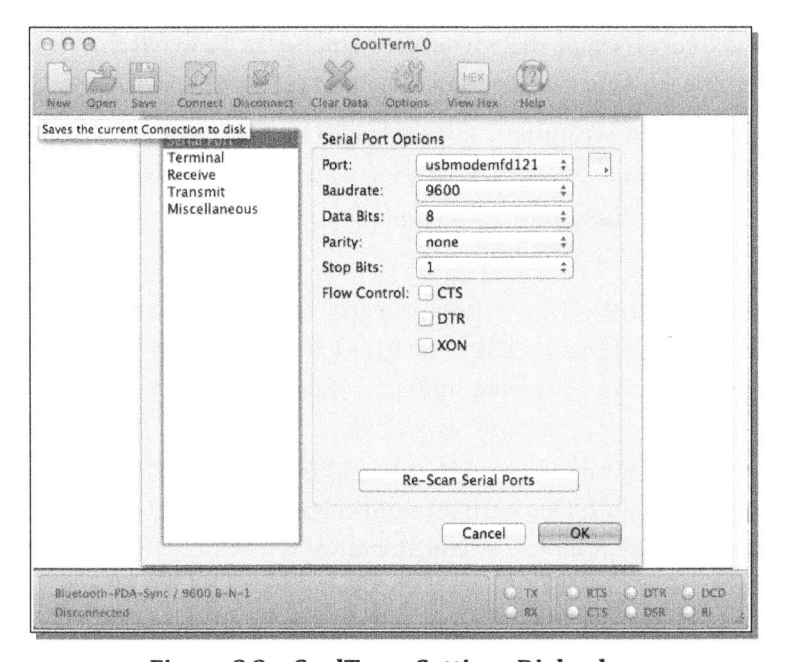

Figure 8.2 – CoolTerm Settings Dialog box

The unique name given to the CDC device, "usbmodemfd121", has been automatically *assigned* by the OS X operating system based on the specific hub/port used to connect. In Windows you will most likely see the new connection reported as a "COMx:" device, where x is a number generally higher than 4 (the original number of hardware serial ports). Note also that the communication parameters typically assigned to a serial port, such as baud-rate, number of bits, parity and number of stop bits can still be modified by the application, but they have *no effect* on the actual operation of the port.

Puts and Gets

On the device side, once the connection is established, a pair of simple (CDC specific) functions will allow us to perform basic serial I/O. The MLA USB library gives us:
- putUSBUSART(buffer, len)
- getsUSBUSART(buffer, len)

These are meant to replace the traditional C library *stdio.h*, puts() and gets(), or the corresponding functions from the PIC24 peripheral library *(uart.h)*, when communicating with the host via the USB link. But there are important differences that we need to highlight:
- Both functions are *non-blocking*: if no data is available, the getsUSBUART() function for example will return immediately with an empty buffer. If the previous putsUSBUART() operation has not completed, issuing a new one will overwrite the buffer and likely corrupt the current transmission. To avoid this, it is necessary to perform a check using the function USBUSARTIsTxTrfReady().
- Both functions can accept only a limited size buffer/string. The receive buffer is limited to 64 characters only (exactly the size of a bulk endpoint), while the transmit buffer is limited to 256 thanks to a slightly more sophisticated buffering performed on the latter.
- For the putsUSBUART() to function properly it is also necessary to perform a periodic call to the function CDCTxService(), which provides the fragmentation and dispatching of the outgoing buffer in 64 bytes sized packets (the size of a bulk endpoint)

Overall these are significant differences that would force any CDC application to be rewritten around a polling main loop with a complex state machine.
But we can make things a bit more manageable if we re-package those two functions and we take full advantage of the USB interrupt hooks provided by the library.

The Callback Handler

Each USB application is requested to define a USB handler callback function. That is a function that will be called by the library every time a special event is detected and gives the application a chance to intervene and perform some additional action on top or in place of the default behavior provided by the library. The event list includes:
- A *Start of Frame* event, which indicates a new USB frame is being sent by the host (once every millisecond, exactly) providing a convenient time base for the application.

- The host requesting the device to enter a Suspend mode (low power stand by)
- A request from the host to Resume from Suspend mode
- A request to provide the descriptors table during enumeration, which gives us an option to modify parts of the tables at run time
- An error on the USB bus

Listing 8.4 illustrate how most of the example applications in the MLA USB library manage such events.

```
BOOL USER_USB_CALLBACK_EVENT_HANDLER(int event, void *pdata, WORD size)
{
    switch( event )
    {
        case EVENT_TRANSFER:
            break;
        case EVENT_SOF:
            USBCB_SOF_Handler();
            break;
        case EVENT_SUSPEND:
            USBCBSuspend();
            break;
        case EVENT_RESUME:
            USBCBWakeFromSuspend();
            break;
        case EVENT_CONFIGURED:
            USBCBInitEP();
            break;
        case EVENT_SET_DESCRIPTOR:
            break;
        case EVENT_EP0_REQUEST:
            USBCBCheckOtherReq();
            break;
        case EVENT_BUS_ERROR:
            break;
        case EVENT_TRANSFER_TERMINATED:
            break;
        default:
            break;
    }
    return TRUE;
}
```

Listing 8.4 - the Event Handlers (call back) function

As we can see the large switch statement vectors individual events to separate callback functions. Not all these functions need to be customized for each application. In fact most of them are simply calling the default handler within the USB library, call a class specific handler or can be left empty.

```
// ************** USB Callback Functions *****************************************
void USBCB_SOF_Handler(void)
{
}

void USBCBSuspend(void)
{
    // Insert appropriate code here to save power ...
    USBSleepOnSuspend();
}

void USBCBWakeFromSuspend(void)
{
    // The host allows 10+ milliseconds of wakeup time,
    // Insert code here to wake up the application ...
}

void USBCBCheckOtherReq(void)
{
    USBCheckCDCRequest();
}

void USBCBInitEP(void)
{
    //Enable the CDC data endpoints
    CDCInitEP();
}
```

Listing 8.5 – Class and application specific Event Handlers

In Listing 8.5 we can see how:

- The Start of Frame callback function is left unused, for now...

- The Suspend call back function is calling the default library function that makes the USB peripheral ready for a sleep state.

- The WakeFromSuspend call back function is left empty

- The class callback functions are calling the CDC class specific default handlers

Using the Start of Frame Handler

As pointed out in the previous section, the Start of Frame handler can provide a convenient time base for an application as it gets called exactly once every millisecond with the beginning of each new frame sent from the host. The timing is extremely accurate, as it is based on the host high precision internal timing resources, but it is provided only when the device is connected.

This is a great resource to use in applications that are powered by the bus (and therefore have no need for timing when disconnected) and in the following we will assume this will be the case of our Mikromedia CDC example project.

As a first application of the start of frame handler, we can use it to ensure that the `CDCTxService()` function is called at least once per frame.

We can also use it to manage the touch screen state machine (see Chapter 4) calling the `TouchDetectPosition()` function at regular intervals.

```
void USBCB_SOF_Handler(void)
{
    static int debounce = 100;

    // service CDC state machine
    if ( USBGetDeviceState() == CONFIGURED_STATE)
        CDCTxService();

    // run the touch screen state machine
    TouchDetectPosition();

    if ( debounce != 0)
        debounce--;

    // debounce screen touches
    if ( TouchGetX() > 0)
    {
        if ( debounce == 0)
        {
            Xvalue = TouchGetX();
            Yvalue = TouchGetY();

            //Wait 100ms before the next press can be generated
            debounce = 100;
        }
    }
} // SOF Handler
```

Listing 8.6 – Start of Frame Handler modified

As we can see from Listing 8.6, we can also use the SOF handler to perform a bit of debouncing logic for us, so that we record a new (X,Y) pair every time the screen is touched but not more often than once every 100ms.

Blocking I/O with the CDC Class

We can now create new blocking puts/gets functions to simplify the use of the CDC class. The goal is to make it easier to port old UART based application into modern USB based application with minimal code modifications.

```c
/*
 * a blocking double buffered puts function
 *
 */
void putsUSBX( char *s)
{
    static char USB_To_Buffer[ 255];

    // ensure connected, active and ready to transmit
    while ( !USBUSARTIsTxTrfReady())
        ;

    // copy string into buffer
    strncpy( USB_To_Buffer, s, 255);

    // send
    putUSBUSART( USB_To_Buffer, strlen( s));
}
```

Listing 8.7 – Blocking putsUSB()

```c
/*
 * a blocking gets function
 */
int getsUSBX( char *s)
{
    int n;

    while ( (n = getsUSBUSART( s, 64)) == 0);

    return n;
}
/*
 * a non blocking gets function
 */
int getsUSB( char *s)
{
    return getsUSBUSART( s, 64);
}
```

Listing 8.8 – Blocking getsUSB()

These routines still require a certain amount of diligence to be used, for example, the string passed to the gets functions must be at least 64 bytes capable but, in return, it is now possible to drop them in the existing flow of a traditional sequential application written for use with a UART without a complete re-write.

A CDC Touch Screen mini Terminal

In an effort of demonstrating the capabilities of the CDC class we will turn the Mikromedia board into a small remote terminal. Once connected to a terminal emulation application on the host, the Mikromedia LCD screen will be used to echo each keystroke entered.
At the same time, since the Mikromedia board does not have a keyboard of its own, we will use the touch screen input to generate messages to send back to the host indicating the screen coordinates of the last touch event.

Let's start the usual process of creating a new project that we will call: "**8-CDC_Serial**". Let's add all the modules required for the Graphics, Touch screen, Serial Flash and LCD terminal emulation as we have done in the previous chapters.
Let's also add a new **USB** logical sub folder in the Source Files section to include the new components: ***usb_device.c***, ***usb_hal_pic24.c*** and ***usb_function_cdc.c***

We can now create a new *main.c* file as in Listing 8.9

```
/*******************************************************************
 Project:   8-CDC_Serial
 File:      main.c

 *******************************************************************/

#include <PICconfig.h>
#include <HardwareProfile.h>

#include <USB/usb.h>
#include <USB/usb_function_cdc.h>
#include <LCDTerminal.h>
#include <TouchScreen.h>
#include <M25P80.h>
#include <uMedia.h>
```

Listing 8.9- main.c – banner and includes

As all the pieces of our first USB application are coming together:

1. We know how to configure our USB connection for use with a CDC class with *usb_config.h* and *usb_descriptors.c*

2. We know how to initialize the USB peripheral and start the enumeration process by calling USBDeviceInit() and USBDeviceAttach() respectively

3. We know how to monitor the enumeration progress and await for the connection to the host to be fully established using USBGetDeviceState()

We can start to sketch out the body of our main() function.

```
/*********************************************************************/
int main(void)
{
    InitializeSystem();

    // Initializes USB module SFRs and firmware
    USBDeviceInit();

    USBDeviceAttach();
    while ( USBGetDeviceState() < CONFIGURED_STATE)
            DisplayUSBStatus();
    DisplayUSBStatus();

    LCDCenterString( 0, "USB CDC Device demo!\n");
    LCDCenterString( 2, "Launch your terminal...\n");

    LCDPutString(">");

    putsUSB( "Hello CDC!\n");

    // main loop
    while( 1)
    {
        DisplayUSBStatus(); // Display Status changes
        ProcessIO();
    }
}//main
```

Listing 8.10 – Main() function

The main loop of our application is now composed of just two items: a call to the `DisplayUSBStatus()` function, so that the user can be alerted in case the connection is lost for some reason, and a call to a `ProcessIO()` function where, borrowing from the template used in most MLA USB demo projects, the actual application logic is implemented.

The code presented in Listing 8.11 is simple enough once we use the *LCDTerminal.c* module and the *Touchscreen.c* module presented in the earlier chapters of this book to define the `InitializeSystem()` and `ProcessIO()` functions.

Obviously we are also responsible for adding all the necessary Mikromedia specific modules (found in the */uMB* folder) that *LCDTerminal.c* and *Touchscreen.c* depend on, including:

- **LCDTerminalFont.c**
- **TouchscreenResistive.c**
- **M25P80.c**
- **drv_spi.c**
- **uMedia.c**

USB - 241

```c
    volatile int Xvalue, Yvalue;

    static void InitializeSystem(void)
    {
        uMBInit();

        LCDInit();
        DisplayBacklightOn();

        TouchInit( NVMWrite, NVMRead, NVMSectorErase, NULL);

    }// InitializeSystem

    void ProcessIO(void)
    {
        int i, n;
        char Send[255];
        char Receive[64];

        if ( Xvalue > 0)
        {
            sprintf( Send, "X = %3d   Y = %3d\r\n", Xvalue, Yvalue );
            putsUSB( Send);
            Xvalue = 0;       // clear flag
        }

        // check if received data
        n = getsUSB( Receive);
        if ( n != 0)
        {
            for( i=0; i<n; i++)
            {
                // print on local terminal
                LCDPutChar( Receive[i]);
                // echo
                Send[i] = Receive[i];
            }
            Send[n] = 0;   // close the string

            // echo back to terminal
            putsUSB( Send);
        }
    }   // ProcessIO
```

Listing 8.11 – InitializeSystem() and ProcessIO() function

CDC Summary

Let's build and program the application on the Mikromedia board. We should be able to test it by following the next few simple steps:

1. Connect the board to the USB port

2. Observe the `DisplayUSBStatus()` function to transition quickly through all the enumeration steps until it says "connected"

3. The application title should appear in the center of the screen followed by a prompt to launch the terminal application on the host

4. If using a Windows machine, let's install the ***mchpcdc.inf*** file provided, otherwise skip this step

5. Launch the terminal emulation application of choice (HyperTerminal, TeraTerm, CoolTerm, MiniTerm...) scan the available serial ports on the system and select the "new" one

6. Establish the connection and verify to receive on the screen the first welcome message from the Mikromedia board: "Hello CDC!"

7. Anything typed on the keyboard should be immediately echoed on the board LCD screen. (Whatch out for terminal CR/LF settings, let's make sure it is set for LF only!)

8. Let's touch the screen of the Mikromedia board and verify that the X, Y coordinates are immediately printed on the computer terminal screen.

If all tests passed, we have just demonstrated bidirectional communication with a host through a virtual communication port using a legacy (terminal) application!

USB-CDC Applications Development Checklist

In summary, the additional steps required to enable a Mikromedia board to use a virtual communication port using the USB/CDC class are:

1. A new logical folder (**USB**) is added to the *Source Files* folder of the project

2. The following items are added to the logical folder:

    - **usb_device.c**, the basic building blocks of any USB device application
    - **usb_hal_pic24.c**, the hardware abstraction layer, .i.e. PIC24 specific support
    - **usb_function_cdc.c**, the CDC class specific support
    - **usb_descriptors.c**, the descriptors tables to be used by the application, copied from the template file provided with the MLA CDC demo projects.

3. Add the following item to the *Include Files* logical folder:

    - **usb_config.h,** the configuration file for the USB library, copied from the template file provided with the MLA CDC demo projects

4. As for all previous MLA projects, configure the compiler *C Include Dirs* to contain:

    - . (dot), the current project directory for MLA to reach our configuration files
    - **../Microchip/Include**, for the source files to reach inside the MLA
    - **../uMB**, for project source files and the MLA to access the Hardware Profile and other resources shared and specific to the Mikromedia board

5. If targeting a Windows host, supply the **mchpcdc.inf** file from the MLA folder **/Device – CDC – Basic Demo/inf**

Human Interface Device Class

Just as in the CDC case, the choice of an HID class model for an application provides some pretty clear guidance on how to configure a USB connection. The standard recipe calls for:

- Endpoint 0: (*Control*) in/out used to enumerate and set up the connection

- Endpoint 1: (*Interrupt*) in/out to move packets of data between device and host

The use of *interrupt* transfers guarantees the integrity of each data packet and at the same time its prompt delivery. This also means that the HID interface is available in Low Speed connections whereas the *bulk* endpoints, used in the CDC class, imply a restriction to use only Full Speed connections or higher.

Just as we did with the CDC class, let's take a look now at a typical *usb_descriptors.c* file that we can borrow from one of the many MLA/HID demo projects and use as a template for our applications.

```
1.      #ifndef __USB_DESCRIPTORS_C
2.      #define __USB_DESCRIPTORS_C
3.
4.      #include "./USB/usb.h"
5.      #include "./USB/usb_function_hid.h"
6.
7.      /* Device Descriptor */
8.      ROM USB_DEVICE_DESCRIPTOR device_dsc=
9.      {
10.         0x12,           // Size of this descriptor in bytes
11.         USB_DESCRIPTOR_DEVICE,  // DEVICE descriptor type
12.         0x0200,         // USB Spec Release Number in BCD format
13.         0x00,           // Class Code
14.         0x00,           // Subclass code
15.         0x00,           // Protocol code
16.         USB_EP0_BUFF_SIZE,  // Max packet size for EP0, see usb_config.h
17.         0x04D8,         // Vendor ID
18.         0x003F,         // Product ID: Custom HID device demo
19.         0x0002,         // Device release number in BCD format
20.         0x01,           // Manufacturer string index
21.         0x02,           // Product string index
22.         0x00,           // Device serial number string index
23.         0x01            // Number of possible configurations
24.     };
```

Listing 8.12 – HID usb_descriptors.c – Device descriptors

The first section of Listing 8.12, shows the main device descriptor.
There are only two elements that are worth noting here:

- Line 17 and 18, contain a *Vendor ID* and *Product ID*. Standard MCHP values are used here, but we might want to edit them later for our own project. See the Tips and Tricks section at the end of this chapter.

- Line 23, shows that we are offering the host only a single configuration choice

```
25.         /* Configuration 1 Descriptor */
26.         ROM BYTE configDescriptor1[]={
27.             /* Configuration Descriptor */
28.             0x09,                              // Size of this descriptor in bytes
29.             USB_DESCRIPTOR_CONFIGURATION,      // CONFIGURATION descriptor type
30.             0x29,0x00,                         // Total length of data for this cfg
31.             1,                                 // Number of interfaces in this cfg
32.             1,                                 // Index value of this configuration
33.             0,                                 // Configuration string index
34.             _DEFAULT | _SELF,                  // Attributes, see usb_device.h
35.             50,                                // Max power consumption (2X mA)
36.
37.             /* Interface Descriptor */
38.             0x09,//sizeof(USB_INTF_DSC),       // Size of this descriptor in bytes
39.             USB_DESCRIPTOR_INTERFACE,          // INTERFACE descriptor type
40.             0,                                 // Interface Number
41.             0,                                 // Alternate Setting Number
42.             2,                                 // Number of endpoints in this intf
43.             HID_INTF,                          // Class code
44.             0,                                 // Subclass code
45.             0,                                 // Protocol code
46.             0,                                 // Interface string index
```

Listing 8.12b – HID usb_descriptors.c – Configuration and Interface descriptors

In the second section, Listing 8.12b, we can see that the configuration is in itself composed of a single interface (line 31) and the immediately following interface descriptor table refers to the HID class (line 43)

```
47.             /* HID Class-Specific Descriptor */
48.             0x09,                              // sizeof(USB_HID_DSC)+3
49.             DSC_HID,                           // HID descriptor type
50.             0x11,0x01,                         // HID Spec Release Number in BCD (1.11)
51.             0x00,                              // Country Code (0x00 for Not supported)
52.             HID_NUM_OF_DSC,                    // Number of class descriptors, see usbcfg.h
53.             DSC_RPT,                           // Report descriptor type
54.             HID_RPT01_SIZE,0x00,               // Size of the report descriptor
55.
56.             /* Endpoint Descriptor */
57.             0x07,                              /* sizeof(USB_EP_DSC)*/
58.             USB_DESCRIPTOR_ENDPOINT,           // Endpoint Descriptor
59.             HID_EP | _EP_IN,                   // EndpointAddress
60.             _INTERRUPT,                        // Attributes
61.             0x40,0x00,                         // size
62.             0x01,                              // Interval
63.
64.             /* Endpoint Descriptor */
65.             0x07,                              /* sizeof(USB_EP_DSC)*/
66.             USB_DESCRIPTOR_ENDPOINT,           // Endpoint Descriptor
67.             HID_EP | _EP_OUT,                  // EndpointAddress
68.             _INTERRUPT,                        // Attributes
69.             0x40,0x00,                         // size
70.             0x01                               // Interval
71.         };
```

Listing 8.12c – HID usb_descriptors.c – HID class specific endpoints definition

In the third section, Listing 8.12c, a new class specific descriptor table is added. This contains details about the HID spec revision in use, r1.11 (line 50) and hints at the presence of a *report* descriptor, a new kind of table that is unique to HID devices. Lines 56 and 64, present respectively the two interrupt endpoints used.

```
72.      //Language code string descriptor
73.      ROM struct{BYTE bLength;BYTE bDscType;WORD string[1];}sd000={
74.      sizeof(sd000),USB_DESCRIPTOR_STRING,{0x0409
75.      }};
76.
77.      //Manufacturer string descriptor
78.      ROM struct{BYTE bLength;BYTE bDscType;WORD string[25];}sd001={
79.      sizeof(sd001),USB_DESCRIPTOR_STRING,
80.      {'M','i','c','r','o','c','h','i','p',' ',
81.      'T','e','c','h','n','o','l','o','g','y',' ',' ','I','n','c','.'
82.      }};
83.
84.      //Product string descriptor
85.      ROM struct{BYTE bLength;BYTE bDscType;WORD string[22];}sd002={
86.      sizeof(sd002),USB_DESCRIPTOR_STRING,
87.      {'S','i','m','p','l','e',' ',' ','H','I','D',' ',
88.      'D','e','v','i','c','e',' ',' ','D','e','m','o'
89.      }};
90.
91.      //Array of configuration descriptors
92.      ROM BYTE *ROM USB_CD_Ptr[]=
93.      {
94.          (ROM BYTE *ROM)&configDescriptor1
95.      };
96.
97.      //Array of string descriptors
98.      ROM BYTE *ROM USB_SD_Ptr[]=
99.      {
100.         (ROM BYTE *ROM)&sd000,
101.         (ROM BYTE *ROM)&sd001,
102.         (ROM BYTE *ROM)&sd002
103.     };
```

Listing 8.12d – HID usb_descriptors.c – Strings tables

The fourth section, in Listing 8.12d, is very similar to the same section in the CDC descriptors file and it presents the strings with the manufacturer name, the product descriptor and the language code. As usual the strings are encoded as Unicode (16-bit) arrays and as such are handled with a certain care.

The last section of Listing 8.12e contains the new report descriptor table.
This is once more a binary encoded structure that offers a huge amount of freedom in specifying how the information in the HID packet is going to be assembled.

```
104.    //Class specific descriptor - HID
105.    ROM struct{BYTE report[HID_RPT01_SIZE];}hid_rpt01={
106.    {
107.        0x06, 0x00, 0xFF,      // Usage Page = 0xFF00 (Vendor Defined Page 1)
108.        0x09, 0x01,            // Usage (Vendor Usage 1)
109.        0xA1, 0x01,            // Collection (Application)
110.        0x19, 0x01,            //     Usage Minimum
111.        0x29, 0x40,            //     Usage Maximum
112.        0x15, 0x01,            //     Logical Minimum
113.        0x25, 0x40,            //     Logical Maximum
114.        0x75, 0x08,            //     Report Size: 8-bit field size
115.        0x95, 0x40,            //     Report Count: Make sixty-four 8-bit fields
116.        0x81, 0x00,            //     Input (Data, Array, Abs):
117.        0x19, 0x01,            //     Usage Minimum
118.        0x29, 0x40,            //     Usage Maximum
119.        0x91, 0x00,            //     Output (Data, Array, Abs)
120.        0xC0}                  // End Collection
121.    };
```

Listing 8.12e – HID usb_descriptors.c – HID report definition

Explaining the complex grammar of such a data structure is beyond the scope of this book. But we can use it as is, trusting that it defines a *collection* (array) of up to 64, 8–bit *values* or, in simpler terms, a very generic buffer of up to 64 bytes.

This practically means that a 64 bytes wide packet of data can be sent from the host to the device (and/or back) during each frame (each millisecond).

As for what that data packet contents are going to mean, this is something that will be decided only between our application on the device and the application running on the host.

Using the HID Class

With the descriptors tables defined, we can now focus on adding the HID functionality to a new project that we will call: ***HID_Simple***

First we need to add to the project Source Files logical folder the HID function support module: **usb_function_hid.c** found in the **/USB/HIDDeviceDriver** source folder of the MLA.

The next steps are identical to the case of the CDC class application:

1. Use the USBDeviceInit() to initialize the USB peripheral module

2. Use the USBDeviceAttach() function to initiate the enumeration sequence

Once more a ***usb_config.h*** file is used to select a number of important library options. Among them:

- Use of *Interrupt mode* by defining the USB_INTERRUPT symbol

- Enable *Full Speed mode*

- Select the on chip *Full Speed transceiver*

- Use the descriptors tables defined in flash by the *usb_descriptors.c* file

We can inspect the contents of a standard HID *usb_config.h* file as provided in one of the examples of the USB/HID class library, for brevity stripped from the verbose comments, in Listing 8.13.

```
#ifndef USBCFG_H
#define USBCFG_H

/** DEFINITIONS ****************************************************/
#define USB_EP0_BUFF_SIZE   8           // Valid Options: 8, 16, 32, or 64 bytes.

#define USB_MAX_NUM_INT     1           // maximum interface number
#define USB_MAX_EP_NUMBER   1           // maximum endpoint number

#define USB_USER_DEVICE_DESCRIPTOR &device_dsc
#define USB_USER_DEVICE_DESCRIPTOR_INCLUDE extern ROM USB_DEVICE_DESCRIPTOR
device_dsc

#define USB_USER_CONFIG_DESCRIPTOR USB_CD_Ptr
#define USB_USER_CONFIG_DESCRIPTOR_INCLUDE extern ROM BYTE *ROM USB_CD_Ptr[]

#define USB_PING_PONG_MODE USB_PING_PONG__FULL_PING_PONG

//-----------------------------------------------------------------------
#define USB_INTERRUPT

//-----------------------------------------------------------------------
#define USB_PULLUP_OPTION USB_PULLUP_ENABLE

#define USB_TRANSCEIVER_OPTION USB_INTERNAL_TRANSCEIVER

#define USB_SPEED_OPTION USB_FULL_SPEED

//-----------------------------------------------------------------------
#define USB_ENABLE_STATUS_STAGE_TIMEOUTS
#define USB_STATUS_STAGE_TIMEOUT     (BYTE)45    //Approximate timeout in ms

//-----------------------------------------------------------------------
#define USB_SUPPORT_DEVICE

#define USB_NUM_STRING_DESCRIPTORS 3

#define USB_ENABLE_ALL_HANDLERS

/** DEVICE CLASS USAGE *********************************************/
#define USB_USE_HID

/** ENDPOINTS ALLOCATION *******************************************/
#define HID_INTF_ID             0x00
#define HID_EP                  1
#define HID_INT_OUT_EP_SIZE     3
#define HID_INT_IN_EP_SIZE      3
#define HID_NUM_OF_DSC          1
#define HID_RPT01_SIZE          28

#endif //USBCFG_H
```

Listing 8.13 – usb_config.h for a simple HID application

Establishing the Connection

Once the process of enumeration is started with the `USBDeviceAttach()` call, we can monitor its progress via the `USBGetDeviceState()` function. No communication should be attempted with the host until we reach the `CONFIGURED_STATE`.

As we did for the CDC example, we can log the status of the connection on the Mikromedia LCD display using the same `DisplayUSBStatus()` function previously introduced in Listing 8.3.

What is dramatically different this time is that there is no (virtual) serial port created on the host side. There is actually very little visible sign of our device being accepted by the host unless we use an (operating system specific) inspection tool.

On a Windows personal computer we would need to open the **Control Panel** and select the **Devices Manager** application. By clicking on the **Human Interface Devices** icon, we would be able to see the list of the presently recognized HID devices.

On a Mac, a similar list can be obtained by selecting the *Apple menu*, in the left top corner of the screen and from there by selecting the **About this Mac** item. Eventually clicking on the **System Report** button in the *About* dialog box, and choosing the *USB* category, will produce a similar list of all USB devices currently recognized by OS X grouped by classes.

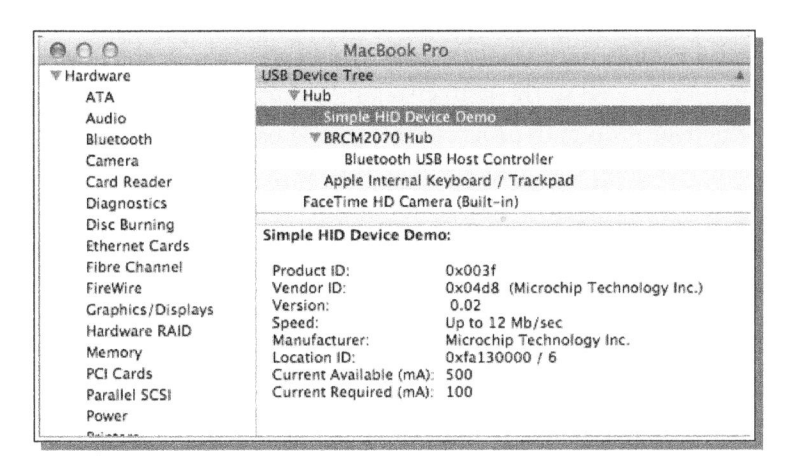

Figure 8.3 - OS X USB Device Tree

HIDTxPacket() and HIDRxPacket()

The MLA HID library gives us a very simple set of functions to play with:

- `HIDTxPacket()`, to deliver a packet of data to the USB SIE for transmission to the host

- `HIDTxHandleBusy()`, to check if the USB SIE is ready to receive new data

- `HIDRxPacket()`, to request a new packet of data

- `HIDRxHandleBusy()`, to check if a new packet of data has been received from the host

Just like the CDC class equivalents, these functions are non-blocking. The way the two pairs of functions interact might need some explanation.

Let's start with the `HIDTxPacket()` function. It takes three parameters:

- An Endpoint (which in the HID case will always be 1)

- A buffer containing the information to be sent, up to 64 bytes

- The amount of data actually contained in the buffer.

The function return value is a *handle*, that is a value that we can use later as the only parameter of the `HIDTxHandleBusy()` function to check if the operation has been completed successfully. Only once the `HIDTxHandleBusy()` returns `FALSE` we can retake control of the buffer we passed to the `HIDTxPacket()` function. Modifying the contents of the buffer while the operation is in progress (busy) means almost certain corruption of the data being transmitted.

The same logic applies to the `HIDRxPacket()` function. It takes similarly three parameters:

- An Endpoint (which in the HID case will always be 1)

- A buffer ready to receive the information

- The size of the receive buffer, up to 64 bytes.

The function return value is a *handle*, that is once more a value that we can use later on as the only parameter of the `HIDRxHandleBusy()` function to check if the receive operation has been completed successfully and fresh new data is available to be collected.

> **NOTE**
>
> Common beginner mistake when using the HID functions is to attempt to read data from the buffer immediately after calling `HIDRxPacket()`. In reality when

> the function returns, the operation might have not even started yet. The data buffer is still "owned" by the USB Serial Interface Engine. Only after the HIDRxHandleBusy() function returns with FALSE, we know that new data has arrived and we get back the ownership of the buffer!

But the order of operations seems to be upside down. When expecting data from the host, we must check first if the receiver is not busy anymore (returns FALSE) and only then we can call the HIDRxPacket() function and access the data!

So there is a chicken and egg kind of problem here. We cannot check the handle status until we have a handle. But we don't get a handle until we call the receive function.

To solve the impasse, there needs to be a first call to the HIDRxPacket() function performed as soon as the HID endpoint is activated at the completion of the enumeration process.

Fortunately the MLA USB library offers us the perfect mechanism to implement this functionality via the *USB Callback Handler* function. This is the same function we used in the CDC demo project in Listing 8.5.

Our HID demo will use a different set of callback functions though (see Listing 8.14):

```
void USBCBSuspend(void){
    USBSleepOnSuspend();
}

void USBCBWakeFromSuspend(void){}

void USBCB_SOF_Handler(void){}

void USBCBCheckOtherReq(void){
    USBCheckHIDRequest();
}

void USBCBInitEP(void)
{
    //enable the HID endpoint
    USBEnableEndpoint(HID_EP,USB_IN_ENABLED|USB_OUT_ENABLED|
                    USB_HANDSHAKE_ENABLED|USB_DISALLOW_SETUP);
    //Re-arm the OUT endpoint for the next packet
    USBOutHandle = HIDRxPacket(HID_EP,(BYTE*)&RxBuffer,64);
}
```

<u>Listing 8.14 – HID Callback functions</u>

Notice that this behavior requires the definition of the Rx and Tx buffers and the respective handles as globals as follows:

```
unsigned char RxBuffer[64];
unsigned char TxBuffer[64];

USB_HANDLE USBOutHandle = 0;    // Must be initialized to 0 at startup.
USB_HANDLE USBInHandle = 0;     // Must be initialized to 0 at startup.
```

<u>Listing 8.15 – HID Buffers and Handles definition</u>

HID Simple Demo

We can now design our main application loop and put the HID class to use.
At first we will re-produce the behavior of the *HID Custom Demo* project provided with the MLA USB library. This will allow us to connect the Mikromedia board to a (Java written) cross platform application called *HID PnP Demo* that is provided with that project.

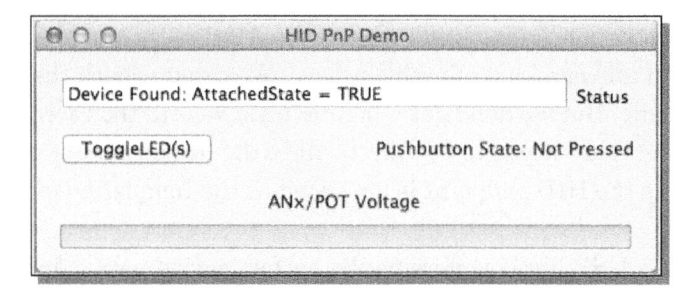

Figure 8.4 – Cross Platform HID Custom Demo Application (OS X)

The application, shown in Figure 8.4, was originally designed to work with the Explorer 16 boards and various USB demo boards, all of which had in common the availability of a small set of LEDs, a pushbutton and a potentiometer.

Once the connection is established, the status bar is updated and with it the other quantities measured by the device. The pushbutton state is reported right under it and the bar at the bottom of the screen is updated continuously to represent the current potentiometer setting. The toggle button allows the user to invert the board LEDs state.
All these operations are controlled by a simple protocol that calls for the host to send commands to the device as the first byte of each new packet. Additional parameters can be passed in the following bytes as required. Some of the commands do require a response data packet to be sent by the device back to the host.
The commands are currently defined as follows:

- 0x80: Toggle LEDs

- 0x81: Get pushbutton state, returns the same code followed by a byte 0/1

- 0x37: Read potentiometer value; returns the same code followed by the ADC reading of the potentiometer channel in two consecutive buffer locations (LSB first)

The protocol is extremely simple if not totally arbitrary, but it represents a good example of many similar protocols used in industrial instrumentation, test and measurement equipment and infinite variations of embedded control systems.

To adapt the demo to the Mikromedia board we will have to perform only a few simple modifications and substitutions:

- We will replace the LEDs with graphical representations on the LCD display (an empty circle will do for the off state, a filled circle will do for the on state)
- We will read the touch screen and report its state instead of the pushbutton.
- We will report the X coordinate of the last touch event instead of the potentiometer value.

To keep with the structure used in the original USB examples we will define a ProcessIO() function that will perform our main loop (only) task. In Listing 8.16 we can see the entirety of the function where we can easily identify the following parts:

- Line 4, we check if the connection is established. If for any reason the application is not in the CONFIGURED state, we cannot call any of the HID packet transfer functions.
- Line 7, we verify if the receive function did complete and new data is available
- Line 11, we select the command based on the content of the first byte in the receive buffer
- Line 14 and 15, we perform the toggle of the LED status.
  This is accomplished using two simple macros that call MLA graphics library drawing primitives.
- Line 18, we process the *report pushbutton* command (0x81) and things get a bit more interesting as we are required to form a response packet.
- Line 20, we check if the transmitter buffer is available, otherwise the operation will have to be ignored for the time being.
- Line 22 to line 27, Only if the previous transmit request has completed we can size the buffer and start composing our reply. It is formed by placing in the first buffer location a copy of the command we are responding to and in the second buffer location the Boolean representation of the TouchScreen current status (TouchGetX() returns -1 when no touch event is detected).
- Line 30, we finally commit the transmit buffer and request the transmission of a packet. Remember when this function returns, the transmission might not have even started!
- Lines 34 to 51, are similarly processing the *report potentiometer* command (0x37) with the only exception that the value returned is the actual X coordinate (scaled

up to form a value in the range between 0 and 1024) and, being a 16-bit value, must be returned in two consecutive locations. Also note that the value is stored in a static variable so the last X coordinate is remembered even after the touch event terminates.

- In Line 58, regardless of which command was received, the ProcessIO function returns after releasing control of the receive buffer to the USB engine re-arming the receive function.

```
1.     void ProcessIO(void)
2.     {
3.         // User Application USB tasks
4.         if ( USBGetDeviceState() < CONFIGURED_STATE) return;
5.
6.         //Check if we have received an OUT data packet from the host
7.         if ( !HIDRxHandleBusy( USBOutHandle))
8.         {
9.             // We just received a packet of data from the USB host.
10.            // Check the first byte of the packet to see what command the host sent
11.            switch( RxBuffer[0])
12.            {
13.                case 0x80:   //Toggle LEDs command
14.                    mLED_1_Toggle();
15.                    mLED_2_Toggle();
16.                    break;
17.
18.                case 0x81:  // Get push button state
19.                    // Ensure the endpoint/buffer is free
20.                    if ( !HIDTxHandleBusy( USBInHandle))
21.                    {
22.                        TxBuffer[0] = 0x81;          // Echo back the command
23.                        if( TouchGetX() >= 0)
24.                            TxBuffer[1] = 0x00;
25.
26.                        else
27.                            TxBuffer[1] = 0x01;
28.
29.                        // Prepare the USB module to send the data packet to the host
30.                        USBInHandle = HIDTxPacket( HID_EP, (BYTE*)&TxBuffer[0], 64);
31.                    }
32.                    break;
```
Listing 8.16 – HID ProcessIO

```
33.            case 0x37:    // Read POT command.  Uses ADC to measure analog voltage
34.            {
35.                static SHORT w;
36.                SHORT t;
37.
38.                // Ensure the endpoint/buffer is available
39.                if ( !HIDTxHandleBusy( USBInHandle))
40.                {
41.                    t = TouchGetX();       // use the X touch position
42.                    if ( t >= 0)
43.                        w = t*3;           // scaled to approx range 1-1000
44.
45.                    TxBuffer[0] = 0x37;    // Echo back to the host the command
46.                    TxBuffer[1] = w;       // LSB
47.                    TxBuffer[2] = w>>8;    // MSB
48.
49.                    //Prepare the USB module to send the data packet to the host
50.                    USBInHandle = HIDTxPacket( HID_EP, (BYTE*)&TxBuffer[0], 64);
51.                }
52.            }
53.            break;
54.        }
55.        //Re-arm the OUT endpoint, so we can receive the next OUT data packet
56.        //that the host may try to send us.
57.        USBOutHandle = HIDRxPacket( HID_EP, (BYTE*)&RxBuffer, 64);
58.    }
59. } // ProcessIO
```

Listing 8.16b - HID ProcessIO (continued)

As we made use of the graphics library, the LCD terminal emulation module and the Touchscreen module, we must take care of initializing them appropriately.

```
void InitializeSystem( void)
{
    DRV_SPI_INIT_DATA si = SPI_FLASH_CONFIG;

    //Init I/Os
    uMBInit();                    // SPI pps, disable analog inputs

    // intialize the Serial Flash
    SST25_CS_LAT = 1;
    SST25_CS_TRIS = 0;
    FlashInit( &si);

    // init display
    LCDInit();
    DisplayBacklightOn();

    TickInit( 1);

    TouchInit( NVMWrite, NVMRead, NVMSectorErase, NULL);
    //TouchInit( NULL, NULL, NULL, NULL);

} // Initialize System
```

Listing 8.17 - HID InitializeSystem() fcuntion

Note that this time we did not use the Start of Frame event handler/interrupt to run the touch screen interface. So, if the Mikromedia board is powered from an alternate source, touch events will be detected even when the device is not connected to the USB bus.

```
int main(void)
{
    InitializeSystem();

    // init USB
    USBDeviceInit();            // Initializes USB module SFRs and firmware
    USBDeviceAttach();

    // wait for USB to connect
    while (USBDeviceState < CONFIGURED_STATE)
        DelayMs(100);
        DisplayUSBStatus();

    // Title and LEDs(!)
    LCDSetColor( BRIGHTRED);
    LCDSetBackground( WHITE);
    LCDClear();
    LCDCenterString( -4, "HID Simple Demo");
    mLED_1_On(); mLED_2_On();

    // main loop
    while(1)
    {
        ProcessIO();
    }
}//end main
```

Listing 8.18 – HID main() function

The `main()` function, in Listing 8.18, can now put together all the bits and pieces of our application providing initialization and the main loop.

A couple of final touches, in Listing 8.19, are still necessary to:

- Provide the main project banner and the list of required `#includes`
- Define the (simulated) LED drawing macros

```
/*******************************************************************
  Project:      8-HID_Simple
  FileName:     main.c
  Hardware:     PIC24 Mikromedia
  Requires:     MLA 1306
 *******************************************************************/

#include "PICconfig.h"

#include <Graphics/Graphics.h>
#include <TouchScreen.h>
#include <M25P80.h>

#include <USB/usb.h>
#include <USB/usb_function_hid.h>

BOOL LED1, LED2;

#define LED1X           90          // X coordinates
#define LED2X           230
#define LEDY            120         // Y coordinates
#define LEDR            30          // radius
#define mGetLED_1()     ( LED1)
#define mGetLED_2()     ( LED2)

#define mLED_1_On()     ( LED1 = 1, SetColor(   RED), FillCircle( LED1X, LEDY, LEDR))
#define mLED_2_On()     ( LED2 = 1, SetColor( GREEN), FillCircle( LED2X, LEDY, LEDR))
#define mLED_1_Off()    ( LED1 = 0, SetColor( WHITE), FillCircle( LED1X, LEDY, LEDR-10))
#define mLED_2_Off()    ( LED2 = 0, SetColor( WHITE), FillCircle( LED2X, LEDY, LEDR-10))

#define mLED_1_Toggle() if ( LED1)   mLED_1_Off(); else   mLED_1_On();
#define mLED_2_Toggle() if ( LED2)   mLED_2_Off(); else   mLED_2_On();
```

Listing 8.19 – Project banner and LED drawing macros

HID Summary

After building the project and programming the Mikromedia board we will be able to:

1. Connect to the host
2. Notice that no driver or .inf file is required even if using a Windows computer
3. Observe the enumeration process as it progresses through the connection states
4. Once reached the *connected* state, let's verify that the application title and the two "LEDs" are shown on the board LCD display
5. Launch the host side application (HID PnP Demo), found for all major platforms in the *USB/Device – HID - Custom Demo* folder of the MLA
6. Test the *HID out* connection, press the Toggle button and observe the LEDs on the Mikromedia screen change state.
7. Test the *HID in* connection, touch the Mikromedia screen. The pushbutton field on the host application will display the instantaneous status. The bottom bar will record the last approximate X coordinate position after each touch event.

HIDAPI

No matter the host operating system used, an HID device is promptly accepted and paired with a standard driver, but there is no established path beyond this point. We need to write both sides of the application from scratch!

This means no legacy support, but at the same time this means a great amount of freedom as we can establish our own protocols for communication between the two sides within the confines of the 64-byte wide packets of data that the HID interface exchanges during each frame.

If on one side we feel a great relief for having the HID interface taking care of the "driver problem", on the other side it might appear as if we have only moved the problem one level up, from the operating system kernel, into the application space.

We might still be very concerned about cross platform compatibility of our application and in general be intimidated by the prospect of having to master Windows, OS X and/or Linux application development.

 The good news is that HID connectivity is very well supported in all major operating systems and in all languages. On a Windows system, for example, we can find support and examples in most all the Visual Studio platforms (C/C++, Java, Visual C#, Visual Basic).

But also most modern *scripting languages* offer libraries that support HID, be it Python, Lua, or Perl.

My personal favorite tool is the ***HIDAPI*** library, an open source library originally developed by Alan Ott.

It offers a common C API across Windows, OS X and Linux and over time has been *wrapped* in many other languages. So you can use HIDAPI calls from Java, Python, or Groovy if you prefer so. No matter the language you are most familiar with or the operating system(s) you target, HID applications are only a few lines of code away.

Rapid Development with Python and HIDAPI

As a final example of the simplicity of using the HID class to develop applications that work across multiple platforms, we will develop a brief application in Python using the HIDAPI library (Python wrapper).

Using the HIDAPI library in Python to connect to the HID Custom Demo requires just two lines of code:

```
import hid
h = hid.device( 0x4d8, 0x3f)
```

Notice that the `hid.device()` function is passed a VID/PID pair (Microchip HID demo pair in this case) to perform a search among all the enumerated devices; if found, it returns a handler to a bidirectional pipe that connects directly the host to the Mikromedia board. Polling the "potentiometer value" for example, can be performed by sending the appropriate command with a simple:

```
h.write( [0x37])
```

Reading the return value can be equally simple:

```
d = h.read(3)
```

This is because both read and write functions are blocking by default although they can be set to became non-blocking by calling:

```
h.set_nonblocking(1)
```

For simplicity, and without requiring any graphics support, we can produce a simple script that polls the HID_Simple application on a connected Mikromedia board and using the sole terminal escape sequences produces a "potentiometer" graph on any host computer screen.

```
#
# graph.py
#
# reads the potentiometer value from the HID Simple demo
#
import hid
import time

# clear screen
print "\x1b[2J\x1b[1;1H"

try:
    h = hid.device(0x4d8, 0x3f)

    while True:
        h.write( [0x37])
        d = h.read(3)
        if d:
            print "*".rjust( (d[1]+d[2]*256) >> 4 )
        time.sleep(0.1)

    print "Closing device"
    h.close()

except IOError, ex:
    print ex

print "Done"
```

Listing 8.20 – HIDAPI Python script

Python comes pre-installed on all Mac and Linux computers, but Windows users have a vast selection of free packages to choose from to install it in a couple of clicks.
Obviously the application can be enriched to include any graphical and animation element as desired by importing additional (Python) libraries such as Tkinter, Qt, or wxPython but this is beyond the scope of this book.

HID Applications Development Checklist

In summary, the additional steps required to enable a Mikromedia board to communicate using the USB/HID class are:

1. Add a new logical folder to the Source Files folder of your project, call it *USB*

2. Add the following items to the logical folder:

    - **usb_device.c**, the basic building blocks of any USB device application
    - **usb_hal_pic24.c**, the hardware abstraction layer, .i.e. PIC24 specific support
    - **usb_function_hid.c**, the HID class specific support
    - **usb_descriptors.c**, the descriptors tables to be used by the application, use the template file provided with this project or from the MLA HID demo projects.

3. Add the following item to the *Include Files* logical folder:

    - **usb_config.h,** the configuration file for the USB library, use the template file provided with this project or from any of the MLA HID demo projects

4. As for all previous MLA projects, configure the compiler **include files path** to contain:

    - . (dot), the current project directory for MLA to reach our configuration files
    - **../Microchip/Include**, for the project source files to reach inside the MLA
    - **../uMB**, for project source files and the MLA to access the Hardware Profile and other resources shared and specific to the Mikromedia board.

Summary

Thanks to the MLA USB library, it is quite simple to connect the Mikromedia board to existing (legacy) applications using the CDC class to create a virtual serial port or using the HID class (and HIDAPI library) to connect with any number of cross platform, scripting tools.

Tips & Tricks

Getting a VID/PID pair

Each USB device regardless of the class it belongs to, must be uniquely identified by a pair of (16-bit) values known as the *Vendor ID* and *Product ID*. As the name implies the Vendor ID is meant to be vendor/company specific and is assigned only by the *USB Implementers Forum* (USB-IF), while the Product ID can be freely defined on a product by product basis. To obtain a Vendor ID, a company or individual needs to register with USB-IF and pay an annual association fee. While this is not a large sum of money, it can be definitely a barrier for the occasional developer, student and hobbyist.

To overcome such limitation, Microchip allows MLA users and developers to share the company VID (0x4D8) using the PID (0x3F) for the purpose of creating demonstration projects.

Further, should there be a need for a small production volume, it is possible to request to sub-license a dedicated unique PID code by applying via email or fax using a form that can be found at: www.microchip.com/usb and selecting the VID/PID FAQs page.

Using multiple classes with Composite Devices

It is possible to define USB applications that span across multiple classes. In consumer products, this is useful to offer a combo device functionality such as a printer and scanner sharing a single USB connection. Among the many MLA USB demo projects you will find a few examples that present themselves to the host computer using special descriptor tables defining such composite devices such as HID and Mass Storage (MSD) or CDC and Mass Storage. These could be used to augment the functionality of our two demo applications by offering sharing of the contents of the Mikromedia microSD card while maintaining a CDC or HID connection.

Suggested Reading

- Jan Axelson – "**USB Complete**" – Lakeview Research
  This is the best and most complete introductory book on USB!

Online Resources

- **USB-IF** – http://www.usb.org

- **HID API** – http://www.signal11.us/oss/hidapi/

- **Python for Windows** – http://www.python.org/getit/windows/

- **Python HIDAPI wrapper** – https://github.com/gbishop/cython-hidap

Exercises

1. Create a simple Python GUI to replicate the HID PnP Demo functionality in twenty lines of code or less

2. Add new commands to the HID protocol to:
   - Control independently the two (simulated) LEDs,
   - Change the display background color
   - Play an audio resource

Alphabetical Index

- AAC 155
- Accelerometer 212
- ADPCM 155
- Audacity 169
- Audio
  - flushMP3() 168
  - Hello.h 171
  - MODE register 158
  - MP3-CS 156
  - MP3-DCS 156
  - MP3-DREQ 156
  - MP3-Reset 156
  - MP3Init() 162
  - readMP3Register() 164
  - setMP3Volume() 169
  - simpleFeedMP3() 167
  - STATUS register 158
  - VOL register 158
  - WAV file header 176
  - writeMP3() 162
  - writeMP3Register() 162
- BacklightInit() 194
- BacklightSet() 195
- Bitmaps 69
- Board Support Package 41
  - ADXL345.c 213
  - adxl345.h 212
  - drv_spi.c 146
  - LCDmenu.c 138
  - LCDTerminal.h 92
  - LCDTerminalFont.c 92
  - m25p80.c 146
  - ScreenCapture() 153
  - TouchScreen.c 100
  - TouchScreenResistive.c 100
  - uMBInit() 126
  - uMedia.c 125
  - VS1053.h 159
- Bootloader
  - Notes 27
  - Programming steps 28
  - Pros & cons 17
- C Includes Dir 57
- codec 155
- CoolTerm 233
- DisplayFadeIn() 63
- EEPROMs 145
- FAT16 121

FAT32 .... **121**
Font ....
    FONTDEFAULT 101
    glyph 79
    TerminalFont 87
Fonts **77**
GDD ....
    GDD_GraphicsConfig.h 204
    GDD_X_Event_Handler.c 204
    GDDDemoGOLDrawCallback() 206
    GDDDemoGOLMsgCallback() 206
    HWP_MIKRO_8PMP.h 205
    InitializeBoard() 206
GDD X **200**
GIMP **93**
GOL **179**
    Blocking 181
    BTN_MSG_PRESSED 188
    BtnCreate() 184
    Color Scheme 197
    Digital Meter 203
    EVENT_INVALID 186
    EVENT_MOVE 186
    GDD_Resource.c 204
    GDD_Screens.c 204
    GetObjID() 186
    GOL_MSG 186
    GOL_SCHEME 197
    GOL.c 191
    GOLCreateScheme() 198
    GOLDefaultColorScheme.c 198
    GOLDeleteObject() 185
    GOLDeleteObjectByID() 185
    GOLDraw() 190
    GOLDrawCallback() 190
    GOLFindObject() 186
    GOLInit() 184
    GOLMsg() 188
    GOLMsgCallback() 188, 198
    Non-Blocking 181
    OBJ_DISABLED 182
    OBJ_DRAW 182
    OBJ_FOCUSED 182
    OBJ_HIDE 182
    Object ID 184
    SldCreate() 196
    SldGetPos() 199
    Style scheme 183
    TextColor0 183

TextColor1.. 183
　　TouchGetMsg().. 187
　　TRANS_MSG.. 188
　　TYPE_TOUCHSCREEN.. 186
　　USE_BUTTON... 193
　　USE_GOL... 193
　　Widget.. 184
　　WndCreate().. 197
Graphic Resource Converter.. 71
Graphics...
　　Bitmap header.. 76
　　ClearDevice()... 53
　　COLOR_DEPTH.. 51
　　Compression.. 73
　　FONT_FLASH.. 78
　　GetMaxX().. 63
　　GetMaxY().. 63
　　GetTextHeight()... 63
　　GetTextWidth().. 63
　　GOLFontDefault.c.. 60
　　Graphics.h... 53
　　GraphicsConfig.h... 193
　　IMAGE_FLASH... 75
　　IMAGE_X2... 71
　　OutText()... 60
　　palette.. 76
　　Primitives... 59
　　PutImage().. 82
　　Resources... 67
　　Run Length Econding... 73
　　SetColor()... 53
　　USE_GOL.. 51
Graphics Config..
　　Creating... 44
　　GDD_GraphicsConfig.h... 204
　　GraphicsConfig.h.. 51, 54, 58, 60, 101, 180, 181, 193, 210
Graphics Display Designer.. 200
Hardware Profile... 44
　　Creating... 45
　　Display section... 49
　　HardwareProfile.h.. 45
　　micro SD / SPI interface section.. 122
　　Serial Flash section.. 147
　　Touch Screen... 101
　　VS1053 section... 160
HIDAPI... 258
ICD3..
　　Programming steps... 28
　　Pros & cons... **18**

Image Decoders..**137**
    BmpDecoder.c... 138
    GifDecoder.c... 138
    ImageAbort().. 139
    ImageDecode()... 139
    ImageDecoderConfig.h... 141
    ImageDecoderInit().. 139
    ImageFullScreenDecode().. 139
    IMG_BMP... 139
    IMG_GIF... 139
    IMG_JPEG... 139
    Jidctint.c.. 138
    JpgDecoder.c.. 138
JTAG..**30**
LAME...**169**
M25P80...**145**
Macros..**32**
MDD File System..**121**
MDD File System..
    FindFirst()... 131
    FindNext().. 131
    FSconfig.c... 123
    FSfeof()... 150
    FSfopen().. 148
    FSfread()... 150
    SearchRec... 131
MIDI..**155**
Mikromedia Board..**16**
Mikromedia Plus..**68**
MikroProg..
    Programming steps... 28
    Pros & cons... 18
MLA..
    Configuring.. 44
    copying.. 41
    Downloading... 37
    Help folder.. 40
    Include folder.. 40
    Installing.. 37
    Microchip folder... 40
    Revisions... 38
MPLAB..
    Code Browsing.. 61
    Code Completion.. 62
    Download.. 9
    Header Files folder... 22
    Important Files folder.. 22
    Installing.. 10
    Library Files Folder.. 22

    Linker Files folder...22
    Loadable Files Folder..22
    Local History..8
    Logical Folders...21
    MPLAB X..7
    New File Wizard..23
    New Project wizard...19
    Source Files Folder..22
        #include...54
MPLAB XC16...**11**
    16-bit Language Tools and Libraries Guide..15
    Compiler User Guide...15
    License Activation Manager..13
    Peripheral Libraries...15
    xc16 folder..14
    XC16 Master Index..15
Ogg Vorbis...**155**
Paint.exe..**93**
Parallel Master Port...**43**
PCM...**155**
PIC24..
    configuration..30
PICconfig.h..**46**
PICKit3...
    Programming steps...28
    Pros & cons...**18**
PPSInput()..**125**
PPSOutput()...**125**
Pygame..**154**
Python..
    for Windows...154
    MP3 converter..170
    Screen Capture...153
    USB HIDAPI...259
QQVGA..**68**
QVGA...**67**
Real ICE...
    Programming steps...28
    Pros & cons...**18**
RIFF..**155**
SD Chip Select..**121**
SD-SPI.c...**121**
Serial Flash..**145**
    FlashInit()..151
    I/O initializations...151
    SF Chip Select..145
    SST25ReadArray()...150
    SST25WriteArray()..150
Sine Test..**164**

T1CON...26
Templates..
    Code 32
    File.. 32
    GDD.. 210
Terminal..
    Emulation... 83
    LCD_BACK... 84
    LCD_FORE.. 84
    LCD_OVERLAY... 89
    LCD_SCROLL.. 84
    LCD_WRAP.. 84
    LCDCenterString()... 91
    LCDClear()... 87
    LCDClearToEOL()... 87
    LCDConfig.h... 84
    LCDInit()... 87
    LCDPut().. 90
    LCDPutChar()... 90
    LCDPutString().. 91
    printf().. 95
    Scrolling.. 83
    write().. 95
TimeDelay.c..49
Timer1...26
Touch..
    Calibration skipping.. 116
    Calibration storing... 151
    Capacitive touch screens... 97
    Resistive touch screens.. 97
    TouchDetectPosition()... 100, 104
    TouchGet()... 110
    TouchGetRawX()... 105
    TouchGetRawY().. 105
    TouchGetX().. 106
    TouchGetY().. 106
    TouchGrid.c.. 111
    TouchGrid()... 110
    TouchHardwareInit().. 105, 106
    TouchInit()...100, 106
    USE_TOUCHSCREEN... 101
uMB folder..41
UNICODE..227
USB..
    ADDRESS_STATE... 231
    ATTACHED_STATE... 231
    Bitrate... 227
    Blocking I/O 238
    cables.. 218

Callback Handler	234
CDC Descriptors	223
Classes	216
Communication Device	222
Composite Devices	262
CONFIGURED_STATE	231
DEFAULT_STATE	231
DETACHED_STATE	231
DisplayUSBStatus()	242
Drivers	217
endpoint	220
Enumeration	221
getsUSBUSART()	234
HID PnP Demo	252
HIDRxHandleBusy()	250
HIDRxPacket()	250
HIDTxHandleBusy()	250
HIDTxPacket()	250
Human Interface Device	222
Human Interface Device Class	244
Introduction	215
J State	219
K state	219
mchpcdc.inf	216
Power	219
POWERED_STATE	231
putUSBUSART()	234
Signaling	218
Start of Frame event	234
Types of Transfer	
Bulk	220
Control	220
Interrupt	220
Isochronous	220
Types of Transfers	220
usb_config.h	229, 248
usb_descriptors.c	223, 244
usb_device.c	222
usb_function_cdc.c	228
usb_function_hid.c	247
usb_hal_pic24.c	222
USB_INTERRUPT	247
usb.h	222
USBDeviceAttach()	247
USBDeviceInit()	247
Virtual Serial Ports	232
VID/PID	262
/USB/Device-CDC-Basic Demo/inf	216
VS1053	**155**

WAV 155
Windows BMP 94
WMA 155
__PIC24FJ256GB110__ 206
_CONFIGx() 47